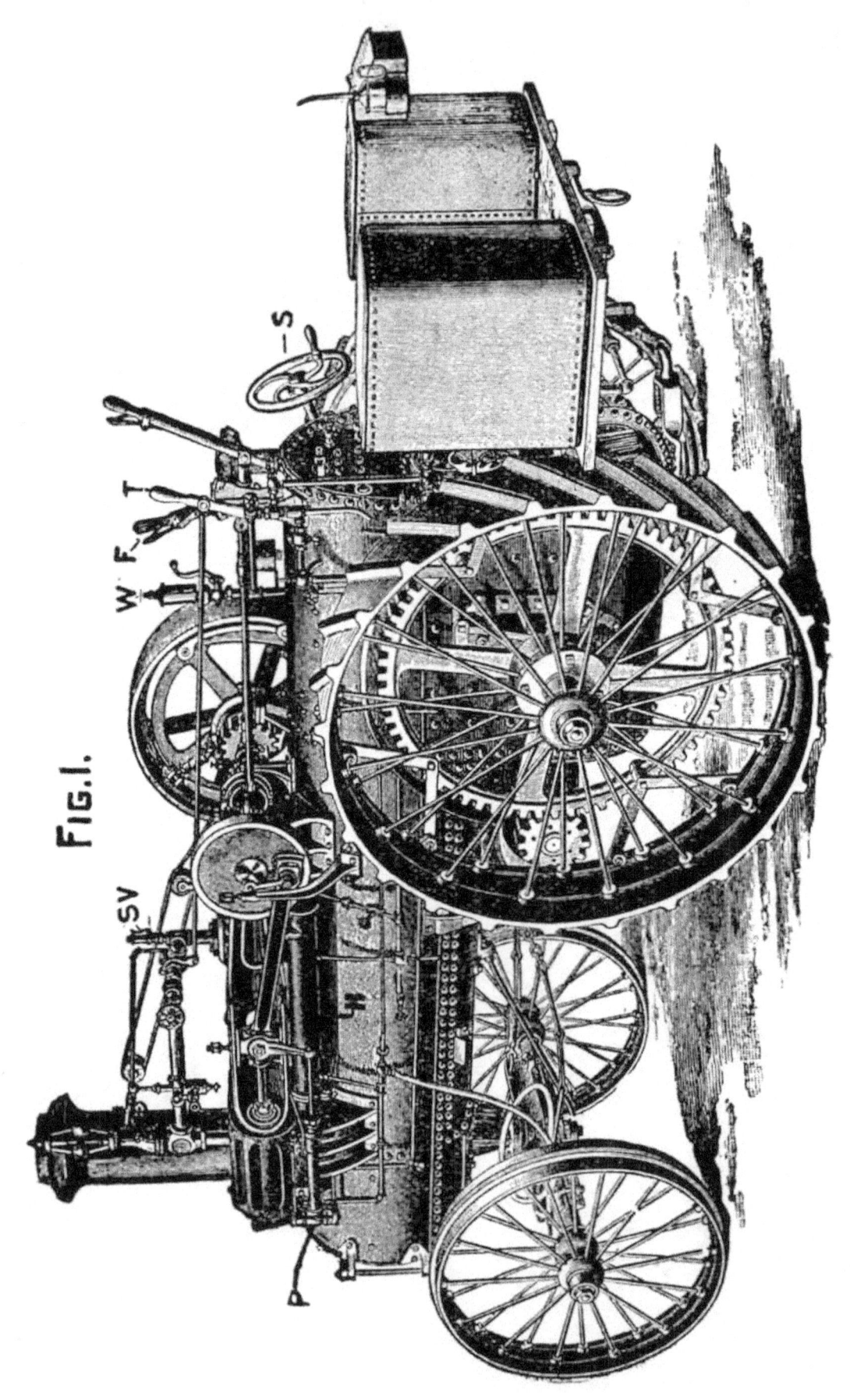

Fig. 1.

A BOOK OF INSTRUCTIONS FOR OPERATORS OF FARM ENGINES

THE

TRACTION ENGINE

ITS USE AND ABUSE.

BY

JAMES H. MAGGARD.

REVISED AND ENLARGED

BY

AN EXPERT ENGINEER.

PHILADELPHIA:
DAVID McKAY, Publisher,
1022 MARKET STREET.
1900.

©2008-2011 Periscope Film LLC
All rights Reserved
ISBN #978-1-935700-56-2

www.PeriscopeFilm.com

Copyright, 1898, by David McKay.

©2008-2011 Periscope Film LLC
All rights Reserved
ISBN #978-1-935700-56-2

www.PeriscopeFilm.com

PREFACE.

In placing this little book before the public the author wishes it understood that it is not his intention to produce a scientific work on engineering. Such a book would be valuable only to engineers of large stationary engines. In a nice engine room nice theories and scientific calculations are practicable. This book is intended for engineers of farm and traction engines—" rough and tumble engineers," who have everything in their favor to-day and to-morrow are in mud holes; who, with the same engine, do eight horse work one day and sixteen horse work next day; who use well water to-day, creek water to-morrow, and water from some stagnant pool next day. Reader, the author has had all these experiences, and you will have them. But do n't get discouraged; you can get through them to your entire satisfaction.

Do n't conclude that all you are to do is to read this book. It will not make an engineer out of you. But read it carefully, use good judgment and common sense, do as it tells you, and, my word for it, in one month you, for all practical purposes, will be a better engineer than four-fifths of the so-called engineers to-day, who think what they do n't know would not make much of a book.

Do n't deceive yourself with the idea that what you get out of this will be merely "book learning." What is said in this will be plain, unvarnished, practical facts. It is not the author's intention to use any scientific terms, but plain, everyday field terms. There will be a number of things you will not find in this book. You will not learn how many pounds of coal it will require to evaporate so many pounds of water, or how many square feet of heating surface is required to produce a given horsepower. You will not find any geometrical figures made up of circles, curves, angles, letters, and figures in a vain effort to make you understand the principle of an eccentric. While it is all very nice to know these things, it is not necessary, and the putting of them in this little work would defeat the very object for which it is intended. Be content with being a good, practical, everyday engineer, and all these things will come in time.

INTRODUCTION.

If you have not read the preface on the preceding pages, turn back and read it. You will see that we have stated there that we will use no scientific terms, but plain everyday talk. It is presumed by us that more young men wishing to become good engineers will read this little work than old engineers. We will, therefore, be all the more plain, and say as little as possible that will tend to confuse the learner, and what we do say will be said in the same language that we would use if we were in the field instructing you how to handle your engine. So if the more experienced engineer thinks we might have gone further in some certain points, he will please remember that by so doing we might confuse the less experienced, and thereby cover up the very point we tried to make. And yet it is not to be supposed that we will endeavor to make an engineer out of a man who never saw an engine. We will, however, insert a description of a traction engine and cuts of some of its parts; not to teach you how to build an engine, but rather how to handle one after it is built—how to know when it is in proper shape and how to let it alone when it is in shape. We will suppose that you already know as much as an ordinary water boy; and just here we will say that we have seen water haulers that were more capable of handling the engine for which they were hauling water than the engineer, and the engineer would not have made a good water boy, for the reason that he was lazy; and we want the reader to stick a pin here, and if he has any symptoms of that complaint, don't undertake to run an engine, for a lazy engineer will spoil a good engine, if by no other means than getting it in the habit of loafing.

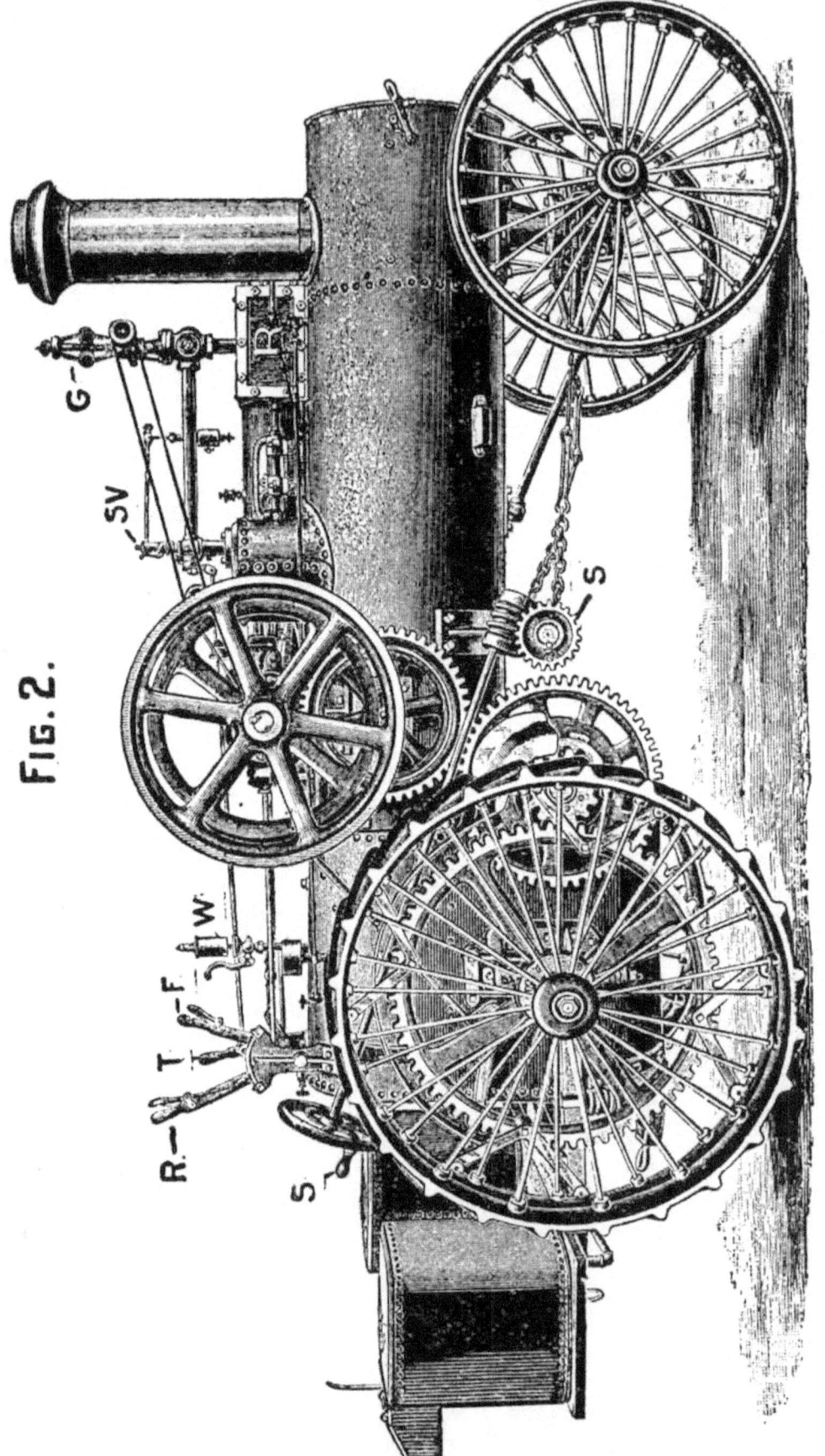

Fig. 2.

PART FIRST.

GENERAL DESCRIPTION OF TRACTION ENGINES.

If you are not already familiar with the principles and parts of such engines you can readily become so by studying figures 1, 2, 3, and 4. Figures 1 and 2 give views of the complete engine and boiler taken from opposite sides. Figure 3 gives the interior construction of a boiler, and figure 4 the section of an engine and accessories.

Look at figure 4. When the throttle-valve on the pipe which connects the boiler with the engine is opened, steam rushes into the valve-chest K. If the engine does not move, it is turned forward a little by hand, which will cause the valve V to move to the left. This motion uncovers the steam passage or port leading into the right-hand end of the cylinder, and the steam rushing into that end of the cylinder pushes the piston P over to the left. Any air or steam in the other end of the cylinder will be driven out through the port into the exhaust, and through the heater into the open air.

The piston will move to the left about one-quarter of its whole motion, the valve also moving in the same direction, but as the valve is operated by a stem driven by an eccentric wheel on the main shaft of the engine, it will, at about this time, commence to move toward the right, closing up the port until when the piston has reached about one-third the distance to the left the valve has closed up over the port and shut off any more steam from going into the right-hand end of the cylinder. The steam will, however, expand and continue to press on the piston, driving it over to the right until it reaches the end of its course. At this time the valve has moved so far to the right that it uncovers the port leading to the left end of the cylinder. Steam rushes in at this end and drives the piston back again to the right, the valve cutting off steam from this end at about one-third of its stroke.

The piston is connected by its rod to the crosshead C, and this crosshead, which moves to and fro under the action of the piston, is connected to the main crank-shaft by means of the connecting rod. By the well-known action of the crank and connecting rod the to-and-fro motion of the crosshead is changed into the rotary motion of the main shaft, which carries the driving wheels.

So much for the engine proper: now for its accessories. It is necessary to change direction or reverse a

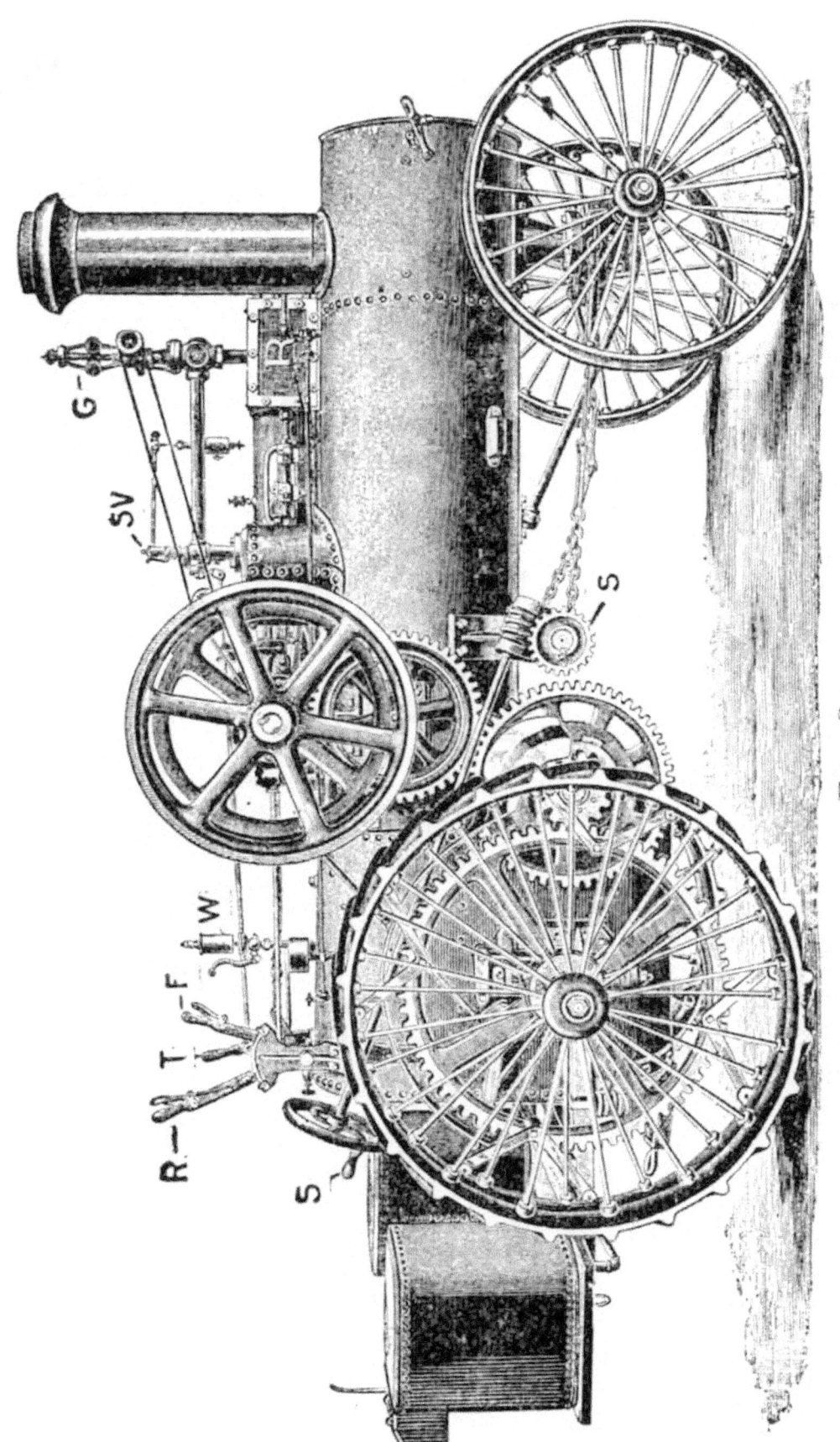

Fig. 3.

6

General Description of Traction Engines.

traction engine, and this is done by moving the reversing lever. In the middle notch the valve is in such a position relatively to the piston that no steam is admitted to the cylinder. Moving the lever to the left moves the eccentric which operates the valve-stem and moves the valve so that steam is admitted to one end of the cylinder, and this will cause the engine to go in one direction. Now, if we throw the reversing lever to the other end of its course, the valve will be moved over so that it admits steam now to the opposite end of the cylinder from which it did before, and the engine will turn in the other direction. In order to stop quickly a brake is provided, operated either by hand or foot lever.

When an engine is propelling itself along the road, the driving wheels turn, of course, much more slowly than the main shaft of the engine. The reduction in speed is obtained by gear wheels. If we want to quickly disconnect the slow-moving road driving wheels so that they do not turn even though the engine shaft is going at full speed, we move the friction clutch-lever F in the proper direction. Sometimes you may come to an obstacle in the road over which the engine refuses to go. You may, perhaps, get over it in this way: Throw the clutch-lever so as to disconnect the road wheels; let the engine get up full speed and then throw the clutch-lever back so as to connect the road wheels.

In figure 1, S is the wheel by which the machine is steered, W is a whistle, and T is the lever controlling the throttle; S V is the safety-valve, and G is the governor.

The pump is shown in figure 4, and is operated directly from the crosshead. It takes water from the water-tank, and pumps it through coils of pipe in the heater to the boiler. The water is heated while passing through the heater, because the pipes through which it flows are surrounded by exhaust steam from the engine.

The general construction of a boiler is shown in figure 3. The flames and hot gases rise from the fuel in the grate and pass through the upper heating chamber, through the tubes, and into the stack. Water fills the cylinder to a level which must be kept above the crown sheet C S, and the heated water gives off steam which collects in the steam dome D, from which it is taken to the engine.

The compound, or two-cylinder, traction engine has come into the market within the last few years, and is the result of trying to secure for farm engines the advantages known and realized for many years by stationary and marine engines. In such engines the steam, after passing through the first or smaller cylinder and expanding somewhat, is exhausted into the second or larger cylinder and allowed to expand completely. Two cylinders are used because we can in this way get better economy in the use of high steam pressure than with the simple engine.

General Description of Traction Engines.

The gain in using high steam pressure can easily be shown:

One hundred pounds of coal will raise a certain quantity of water from 60 degrees into steam at 5 pounds pressure; 102.9 pounds will raise it to 80 pounds; 104.4 pounds will raise it to 160 pounds. That is, by burning $1\frac{1}{2}$ pounds more coal than we used for 80 pounds, we can raise it to 160 pounds, and this steam at 160 pounds run into the engine would give a large increase in power over what we had at 80 pounds for a trifling increase in coal burned. These engines will furnish the same number of horse power with considerably less fuel than simple engines, from 15 to 30 per cent. less, but only when they are run at nearly full load all the time.

If they are to be used on such service as to be lightly loaded for a considerable part of the day, instead of saving coal, as compared with simple engines, they will waste it.

The increased danger from the use of high pressure steam—150 pounds—is counterbalanced by making the boilers stronger than usual in the same proportion as the increase in pressure. Figure 5 shows a compound traction engine which, you will see, differs but little in general appearance from the simple engine.

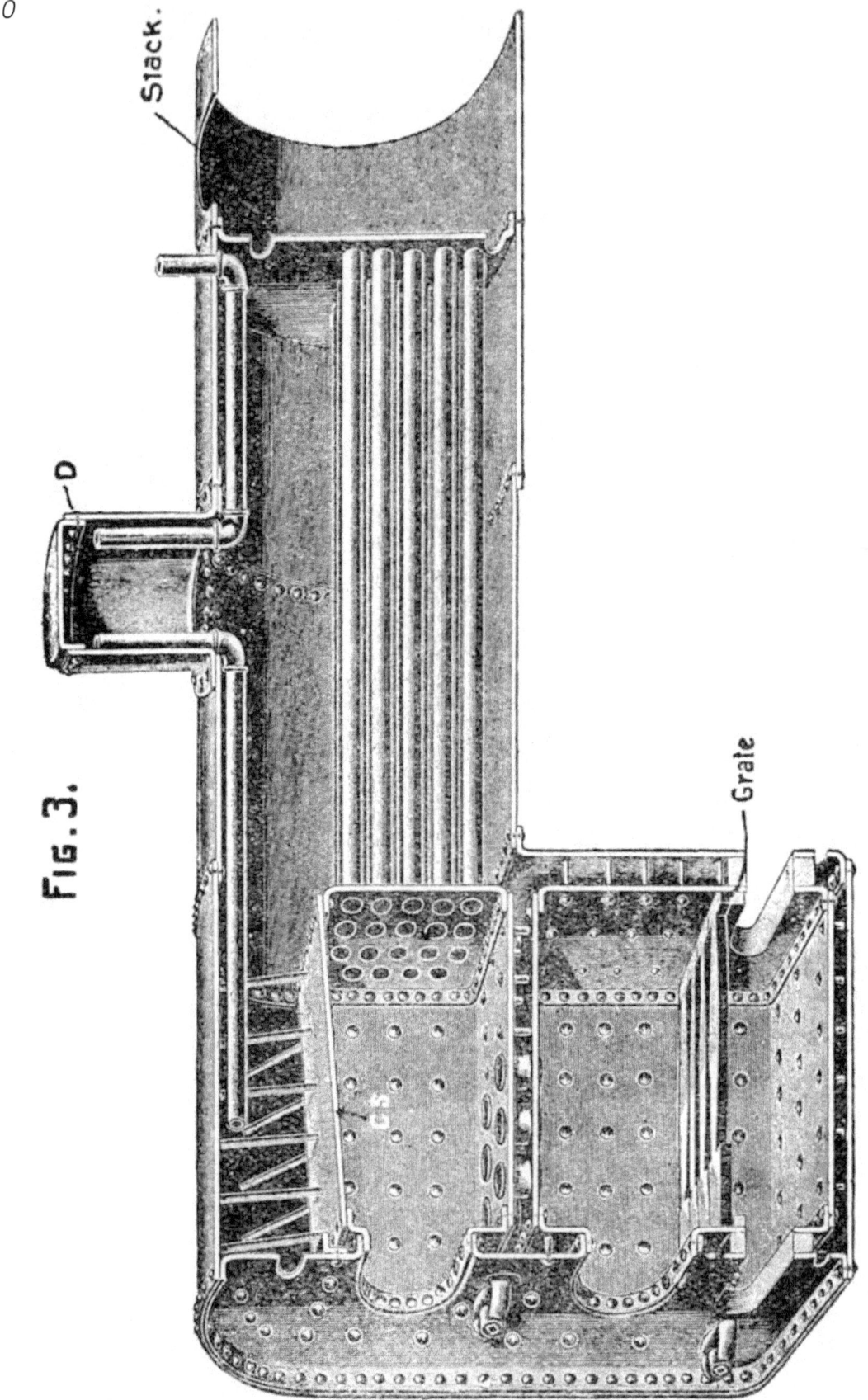

PART SECOND.

WHAT TO DO AND WHAT NOT TO DO.

In order to get the learner started, it is reasonable to suppose that the engine he is to run is in good running order. It would not be fair to put the green boy on to an old dilapidated, worn-out engine, for he might have to learn too fast in order to get the engine to running in good shape. He might have to learn so fast that he would get the big head, or have no head at all, by the time he got through with it. And I do n't know but that a boy without a head is about as good as an engineer with a big head. We will, therefore, suppose that his engine is in good running order. By good running order we mean that it is all there and in its proper place, and that with from ten to twenty pounds of steam, the engine will start off at a good lively pace. And let us say here (remember that we are talking of the lone engine, no load considered) that if you are starting a new engine and it starts off nice and easy with twenty pounds, you can make up your mind that you have an engine that is going to be nice to handle and give you but little, if any, trouble. But if it should require fifty or sixty pounds to start it, you want to keep your eyes open, something is tight; but don't take it to pieces. You might get more pieces than you would know what to do with. Oil the bearings freely and put your engine in motion, and run it carefully for a while and see if you don't find something getting warm. If you do, stop and loosen up a very little and start it up again. If it still heats, loosen about the same as before, and you will find that it will soon be all right. But remember to loosen but very little at a time, for a box or journal will heat from being too loose as quickly as from being too tight, and you will make trouble for yourself, for, inexperienced as you are, you don't know whether it is too loose or too tight, and if you have found a warm box, don't let that box take all of your attention, but keep an eye on all other bearings. Remember that we are not threshing yet; we just run the engine out of the shed (and for the sake of the engine and the young engineer we hope that it did not stand out all winter) and are getting in shape for a good fall's run. In the meantime, to find out if anything heats, you can try your pumps; but to help you along we will suppose that your pump, or injector, as the case may be, works all right.

What to Do and What Not to Do.

Now, suppose we go back where we started this new engine, that was slow to start with less than fifty pounds, and when it did start we watched it carefully and found after oiling thoroughly that nothing heated, as far as we could see. So we conclude that the trouble must be in the cylinder. Well, what next? Must we take off the cylinder head and look for the trouble? Oh, no, not by any means. The trouble is not serious. The rings are a little tight, which is no serious fault. Keep them well oiled, and in a day or two ten pounds will start the empty engine in good shape. If you are starting an engine that has been run, the above instructions are not necessary, but if it is a new one, these precautions are not out of the way, and a great deal of the trouble caused in starting a new engine can be avoided if these precautions are observed.

It is not uncommon for a hot box to be caused from a coal cinder dropping in the box in shipment, and before starting a new engine clean out the boxes thoroughly, which can be done by taking off the caps, or top box, and wiping the journal clean with an oily rag or waste, and every engineer should supply himself with this very necessary article, especially if he is the kind of an engineer who intends to keep his engine clean.

The engine should be run slowly and carefully for a while, to give a chance to find out if anything is going to heat, before putting on any load.

Now, if your engine is all right, you can run the pressure up to the point of blowing off, which is from 100 to 110 pounds. Most new pop-valves, or safety-valves, are set at this pressure. I would advise you to fire to this point, to see that your safety is all right. It is not uncommon for a new pop to stick, and as the steam runs up it is well to try it by pulling the relief lever. If, on letting it go, it stops the escaping steam at once, it is all right. If, however, the steam continues to escape, the valve sticks in the chamber. Usually, a slight tap with a wrench or a hammer will stop it at once, but never get excited over escaping steam, and perhaps this is as good a place as any to say to you, don't get excited over anything. So long as you have plenty of water, and *know* you have, there is no danger.

The young engineer will most likely wonder why we have not said something about the danger of explosions. We did not start out to write about explosions. That is just what we don't want to have anything to do with. But, you say, is there no danger of a boiler explosion? Yes; but if you wish to explode your boiler you must treat it very differently from the way we advise. We have just stated that so long as you have plenty of water, and *know* you have, there is no danger. Well, how are you to know? This is not a difficult thing to know, provided your boiler is fitted with the proper appliances, and

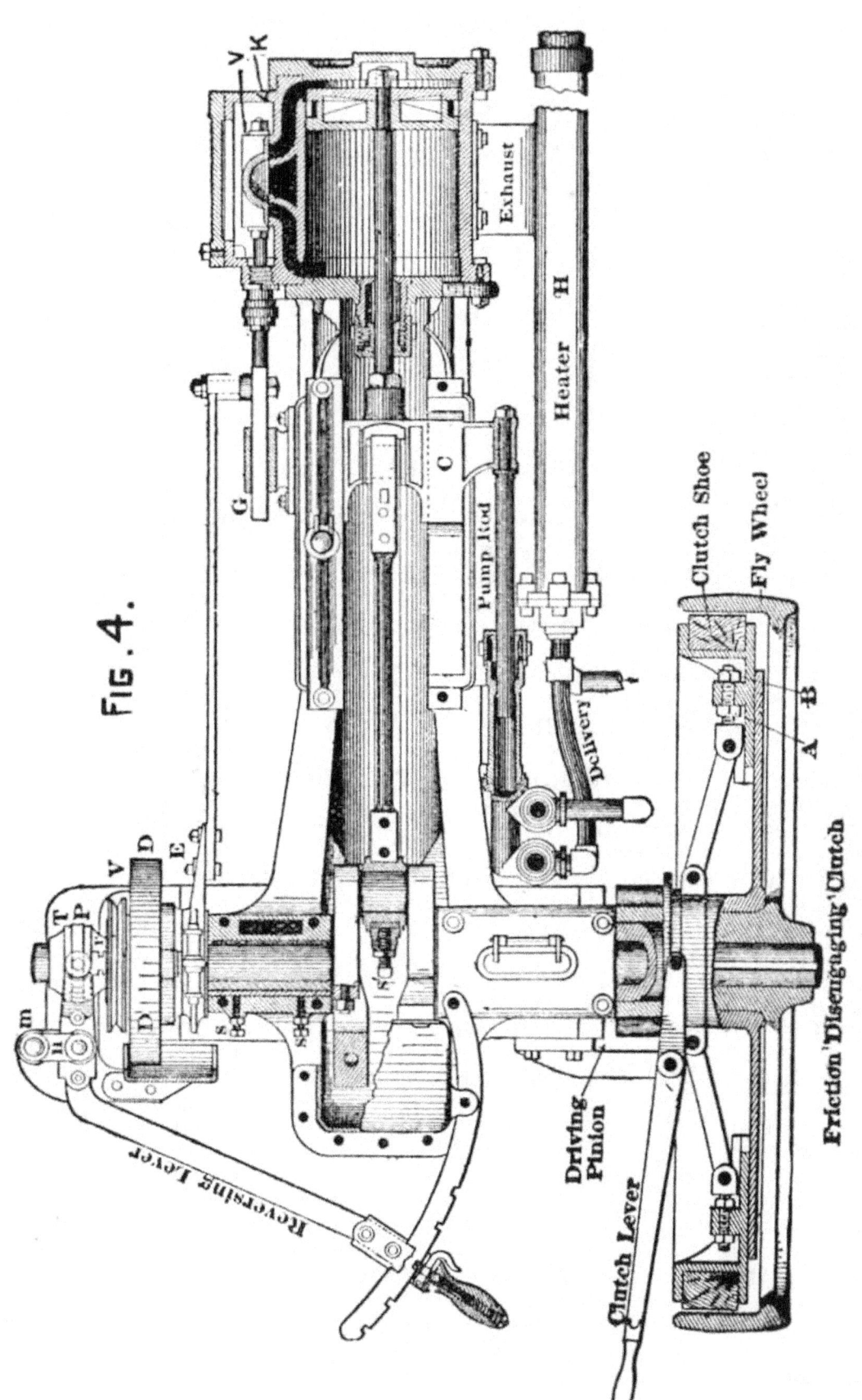

Fig. 4.

all builders of any prominence, at this date, fit their boilers with from two to four try-cocks and a glass gauge. The boiler is tapped in from two to four places for the try-cocks, the location of the cocks ranging from a line on a level with the crown sheet, or top of fire-box, to eight inches above, depending somewhat on the amount of water space above the crown sheet, as this space differs very materially in different makes in the same sized boiler. The boiler is also tapped on or near the level of crown sheet, to receive the lower water glass cock and directly above this, for the top cock. The space between this shows the safe variation of the water. Do n't let the water get above the top of the glass, for if you are running your engine at hard work, you may knock out a cylinder head, and do n't let it get below the lower gauge, or you may get your own head knocked off.

The bottom of the glass gauge is just a little above the crown sheet of the fire-box. So long as this is covered with water it will not get too hot to do any hurt, but if there is n't water enough to cover it, the heat will twist it and cause an explosion, perhaps, unless it is provided with a fusible plug.

Now, the glass gauge is put on for your convenience, as you can determine the location of the water as correctly by this as if you were looking directly into the boiler, provided the glass gauge is in perfect order. But

What to Do and What Not to Do.

as there are a number of ways in which it may become disarranged, or unreliable, we want to impress on your mind that you must not depend on it entirely. We will give these causes further on. You are not only provided with the glass gauge, but with the try-cocks. These cocks are located so that the upper and lower cock is on, or near, the level with the lower and upper end of glass gauge. With another try-cock about on a level with the center of glass gauge, or, in other words, if the water stands about the center of glass, it will at the same time show at the cock when tried. Now, we will suppose that your glass gauge is in perfect condition and the water shows two inches in the glass. You now try the lower cock, and find plenty of water; you will then try the next upper cock and get only steam. Now, as the lower cock is located below the water line, shown by the glass, and the second cock above this line, you not only see the water line by the glass, but you have a way of proving it. Should the water be within two inches of the top of the glass, you again have the line between two cocks and can also prove it. Now you can know for a certainty where the water stands in the boiler, and we repeat when you *know* this, there is nothing to fear from this source; and as a properly constructed boiler never explodes, except from low water or high pressure, and as we have already cautioned you about your safety-valve, you have nothing to fear, provided you have made up your mind to follow these instructions, and unless you can do this let your job to one who can. Well, you say you will do as we have directed ; we will then go back to the gauges. Do n't depend on your glass gauge alone, for several reasons. One is, if you depend on the glass entirely, the try-cocks become limed up and are useless, solely because they are not used.

Some time ago I was standing near a traction engine when the engineer (I guess I must call him that) asked me to stay with the engine a few minutes. I consented. After he had been gone a short time I thought I would look after the water. It showed about two inches in the glass, which was all right, but as I have advised you, I proposed to know that it was there, and thought I would prove it by trying the cocks. But on attempting to try them I found them limed up solid. Had I been hunting for an engineer, that fellow would not have secured the job. Suppose that before I had looked at the glass it had bursted, which it is liable to do any time. I would have shut the gauge-cocks off as soon as possible, to stop the escaping steam and water. Then I would have tried the cocks to find where the water was in the boiler. I would have been in a bad boat, not knowing whether I had water or not. Shortly after this the fellow that was helping the engine to run (I guess I will put it that way) came back. I asked him what the trouble was with his gauge-cocks. He said, "Oh, I don't bother with them." I asked him what he would do if his glass should break. His reply was, "Oh, that won't break." Now, just such an engineer as that spoils many a good engine, and then blames it on the manufacturer. Now, this is one good reason why you are not to depend entirely on the glass gauge. Another equally as good reason is that your glass may fool you, for you see the try-cocks may lime up; so may your glass gauge-cocks, but you say you use them. You use them by looking at them. You are not letting the steam or water escape from them every few minutes, and thereby cutting the lime away, as is the case with try-cocks. Now, you want to know how you are to keep them open. Well, that is easy. Shut off the top gauge and open the drain-cock at bottom of gauge-cock. This allows the water and steam to flow out of the lower cock; then, after allowing it to escape a few seconds, shut off the lower gauge and open the top one, and allow it to blow about the same time; then shut the drain-cock and open both gauge-cocks, and you will see the water seek its level, and you can rest assured that it is reliable. This little operation I want you to perform every day you run an engine. It will prevent you from thinking you have water. I don't want you to think so; I intend that you shall *know* it. You remember we said, if you *know* you

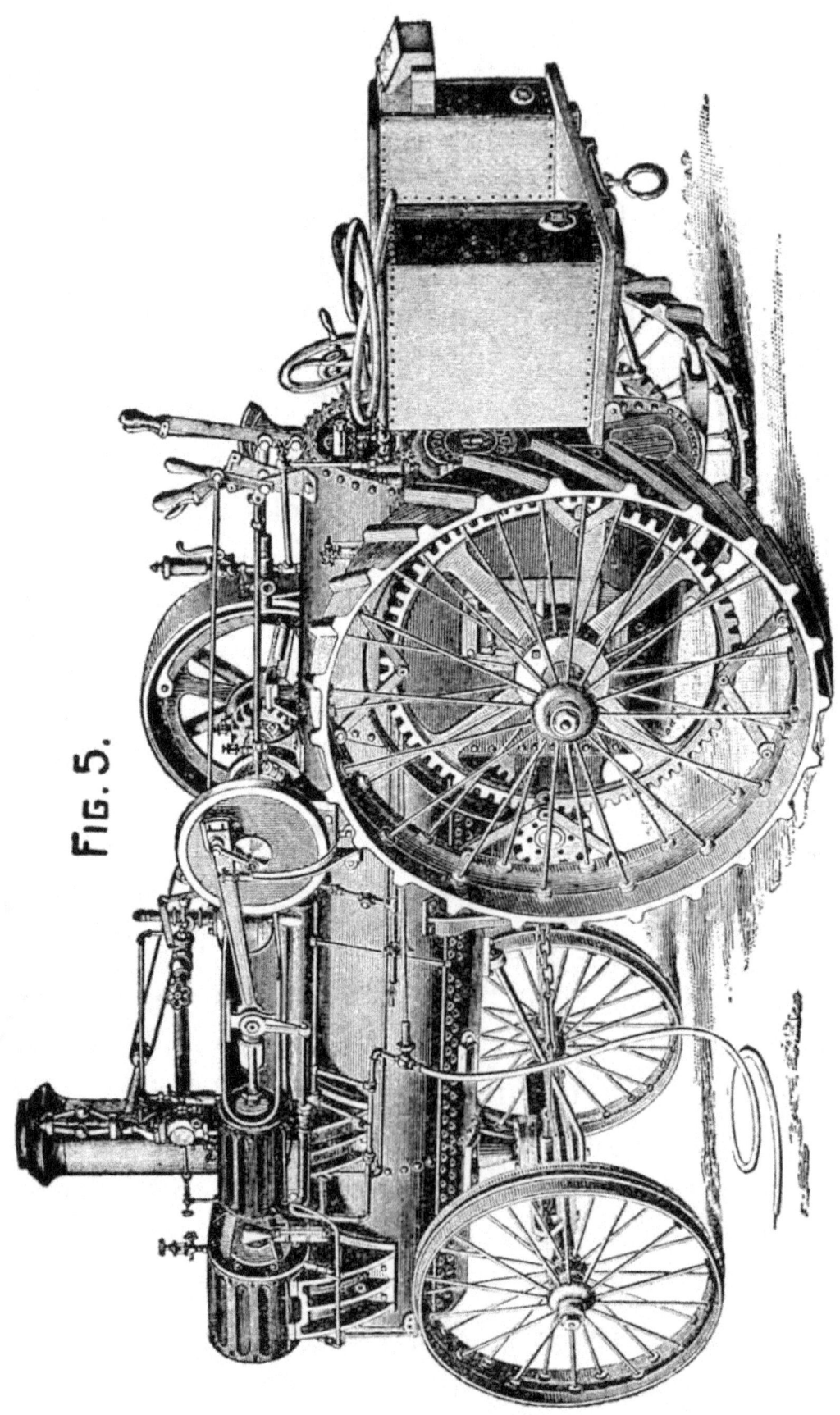

Fig. 5.

have water you are safe, and every one around you will be safe.

Should the gauge-glass break, shut both the gauge-glass valves, and loosen up the lock nuts at top and bottom of the glass; take out the old and put in the new glass, tightening up the lock nuts; then open the valves and test by the try-cocks whether the glass gauge shows the right height of water.

Now, here is something I want you to remember. Never be guilty of going to your engine in the morning and building a fire simply because you see water in the glass. We could give you the names of a score of men who have ruined their engines by doing this very thing. You, as a matter of course, want to know why this can do any harm. It could not, if the water in the boiler was as high as it shows in the glass, but it is not always there, and that is what causes the trouble. Well, if it showed in the glass, why was it not there? You probably have lived long enough in the world to know that there are a great many boys in it, and it seems to be second nature with them to turn everything on an engine that it is possible to turn. All glass gauge-cocks are fitted with a small hand wheel. The small boy sees this about the first thing and he begins to turn it, and he generally turns as long as it turns easy, and when it stops he will try the other one, and when it stops he has done

WATER GAUGES.

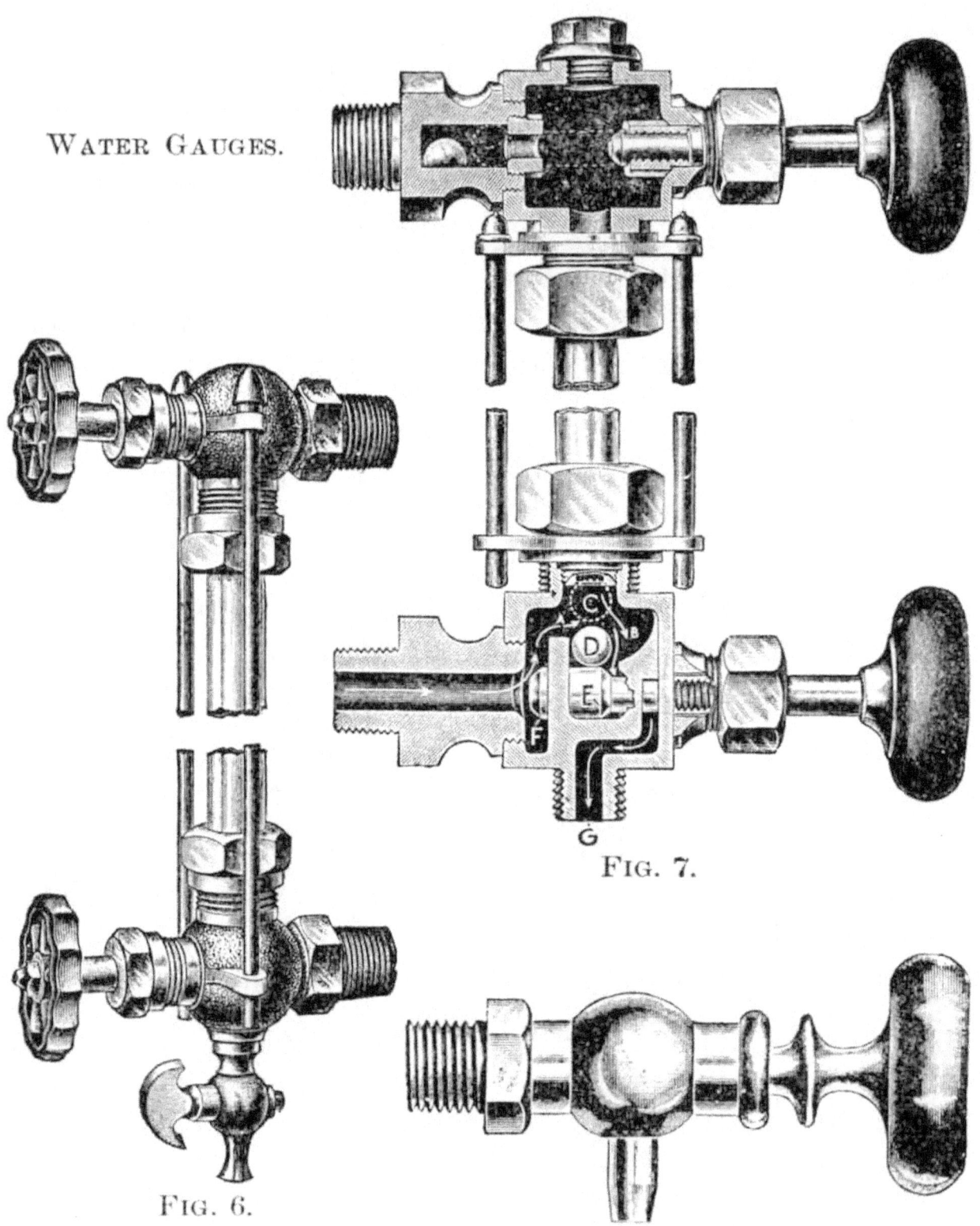

FIG. 6.

FIG. 7.

FIG. 8.—TRY-COCK.

the mischief, by shutting the water off from the boiler, and all the water that was in the glass remains there. You may have stopped work with an ordinary gauge of water, and, as water expands when heated, it also contracts when it becomes cool. Water will also simmer away, if there is any fire left in the fire-box, especially if there should be any vent or leak in the boiler, and the water may by morning have dropped to as much as an inch below the crown sheet. You approach the engine and, on looking at the glass, see two or three inches of water. Should you start a fire without investigating any further, you will have done the damage, while if you try the gauge-cocks first you will discover that some one has tampered with the engine. The boy did the mischief through no malicious motives, but we regret to say that there are people in this world who are mean enough to do this very thing, and not stop at what the boy did unconsciously, but, after shutting the water in the gauge for the purpose of deceiving you, they then go to the blow-off cock and let enough water out to insure a dry crown sheet. While I detest a human being guilty of such a dastardly trick, I have no sympathy to waste on an engineer who can be caught in this way. So, if by this time you have made up your mind never to build a fire until you know where the water is, you will never be fooled and will never have to explain an accident by saying, " I thought I had plenty

What to Do and What Not to Do.

of water." A good authority on steam boilers says: "All explosions come either from poor material, poor workmanship, too high pressure, or a too low gauge of water." Now, to protect yourself from the first two causes buy your engine from some factory having a reputation for doing good work and for using good material. The last two causes depend very much on yourself, if you are running your own engine. If not, then see that you have an engineer who knows when his safety-valve is in good shape, and who knows when he has plenty of water, or knows enough to pull his fire when, for some reason, the water should become low. If poor material and poor workmanship were unknown, and carelessness in engineers were unknown, such a thing as a boiler explosion would also be unknown.

You no doubt have made up your mind by this time that I have no use for a careless engineer, and let me add right here that if you are inclined to be careless or forgetful (they both mean about the same thing), you are a mighty poor risk for an insurance company; but, on the other hand, if you are careful and attentive to business, you are as safe a risk as any one, and your success and the durability and life of your engine depend entirely upon you, and it is not worth your while to try to shift the responsibility of an accident to your engine on to some one else.

If you should go away from your engine and leave it with the water boy, or any one who might be handy, or leave it alone, as is often done, and something goes wrong with the engine, you are at fault. You had no business to leave it; but, you say, you had to go to the separator and help fix something there. At the separator is not your place. It is not our intention to tell you how to run both ends of an outfit. We could not tell you if we wanted to. If the men at the separator can't handle it, get some one who can. Your place is at the engine. If your engine is running nicely, there is all the more reason why you should stay by it, and that is the way to keep it running nicely. I have seen twenty dollars' damage done to the separator and two days' time lost, all because the engineer was as near the separator as he was to the engine when a root went into the cylinder. Stay with your engine, and if anything goes wrong at the separator you are ready to stop and stop quickly, and if you are signalled to start you are ready to start at once. You are, therefore, making time for your employer or for yourself, and to make time while running a threshing outfit means to make money. There are engineers running engines to-day who waste time enough every day to pay their wages.

There is one thing that may be a little difficult to learn, and that is to let your engine alone when it is all right. I once gave a young fellow a recommendation to a farmer

What to Do and What Not to Do.

who wanted an engineer, and afterward noticed that whenever I happened around he immediately picked up a wrench and commenced to loosen up, first one thing and then another. If that engineer ever loses that recommendation he will be out of a job, if his getting one depends on my giving him another. I wish to say to the learner that that is not the way to run an engine. Whenever I happen to go around an engine—and I never lose an opportunity—and see an engineer watching his engine (now don't understand me to mean standing and gazing at it), I conclude that he knows his business. What I mean by watching an engine is, every few minutes let your eye wander over the engine and you will be surprised to see how quickly you will detect anything out of place. So, when I see an engineer watching his engine closely while running, I am most certain to see another commendable feature in a good engineer, and that is, when he stops his engine he will pick up a greasy rag and go over his engine carefully, wiping every working part, watching or looking carefully at every point that he touches. If a nut is working loose, he finds it; if a bearing is hot, he finds it; if any part of his engine has been cutting, he finds it. He picked up a greasy rag instead of a wrench, for the engineer that understands his business and attends to it never picks up a wrench unless he has something to do with it. The good engineer took a greasy rag, and while he was using it to clean his engine, he was at the same time carefully examining every part. His main object was to see that everything was all right. If he had found a nut loose or any part out of place, then he would have taken his wrench, for he had use for it.

Now, what a contrast there is between this engineer and a poor one; and, unfortunately, there are hundreds of poor engineers running portable and traction engines. You will find a poor engineer very willing to talk. This is bad habit number one. He can not talk and have his mind on his work. Beginners must not forget this. When I tell you how to fire an engine, you will understand how important it is. The poor engineer is very apt to ask an outsider to stay at his engine while he goes to the separator to talk. This is bad habit number two. Even if the outsider is a good engineer, he does not know whether the pump is throwing more water than is being used or whether it is throwing less. He can only ascertain this by watching the column of water in the glass, and he hardly knows whether to throw in fuel or not. He does n't want the steam to go down and he does n't know at what pressure the pop-valve will blow off. There may be a box or journal that has been giving the engineer trouble, and the outsider knows nothing about it. There are a dozen other good reasons why bad habit number two is very bad.

What to Do and What Not to Do.

If you will watch the poor engineer when he stops his engine, he will, if he does anything, pick up a wrench, go around to the wrist-pin, strike the key a little crack, draw a nut or peck away at something else, and can't see anything for grease and dirt. When he starts up again, ten to one the wrist-pin heats, and he stops and loosens it up, and then it knocks. Now, if he had picked up a rag instead of a wrench he would not have hit that key, but he would have run his hand over it, and if he had found it all right he would have let it alone, and would have gone over the balance of the engine; and when he started up again his engine would have looked better for the wiping it got, and would have run just as well as before he stopped it. Now, you will understand why a good engineer wears out more rags than wrenches, while a poor one wears out more wrenches than rags. Never bother an engine until it bothers you. If you do, you will make lots of grief for yourself.

I have mentioned the bad habits of a poor engineer so that you may avoid them. If you carefully avoid all the bad habits connected with the running of an engine, you will be certain to fall into good habits and will become a good engineer.

After carelessness, meddling with an engine comes next in the list of bad habits. The tinkering engineer never knows whether his engine is in good shape or not, and the chances are that if he should get it in good shape he would not know enough to let it alone. If anything does actually get wrong with your engine, do not be afraid to take hold of it, for something must be done, and you are the one to do it; but before you do anything, be certain that you know what is wrong. For instance, should the valve become disarranged on the valve-stem, or in any other way, do not try to remedy the trouble by changing the eccentric, or, if the eccentric slips, do not go to the valve to mend the trouble. I am well aware that among young engineers the impression prevails that a valve is a wonderful piece of mechanism, liable to kick out of place and play smash generally. Now, let me tell you right here that a valve (I mean the ordinary slide-valve such as is used on traction and portable engines) is one of the simplest parts of an engine, and you are not to lose any sleep about it, so please be patient until I am ready to introduce you to this part of your work. You have a perfect right to know what is wrong with the engine. The trouble may not be serious, and yet it is important to you that the engine is not running just as nicely as it should. Now, if your engine runs irregularly,—that is, if it runs up to a higher speed than you want and then runs down,—you are likely to say at once: "Oh, I know what the trouble is, it is the governor." Well, suppose it is, what are you going to do about it? Are you going to

What to Do and What Not to Do.

shut down at once and go to tinkering with it? No, don't do that; stay close to the throttle-valve and watch the governor closely. Keep your eye on the governor stem, and when the engine starts off on one of its high-speed tilts you will see the stem go down through the stuffing-box and then stop and stick in one place until the engine slows down below its regular speed, and it then lets loose and goes up quickly, and your engine lopes off again. You have now located the trouble. It is in the stuffing-box around the little brass rod or governor stem. The packing has become dry, and by loosening it up and applying oil you may remedy the trouble until such time as you can repack it with fresh packing. Candle-wick will do for this purpose until regular packing can be obtained.

But if the governor does not act as I have described, and the stem seems to be perfectly free and easy in the box, and the governor still acts queerly, starting off and running fast for a few seconds, and then suddenly concluding to take it easy, and away goes the engine again, see if the governor belt is all right, and if it is, it would be well for you to stop and see if a wheel is not loose. It might be either the little belt-wheel or one of the little cog-wheels. If you find these are all right, examine the spool on the crank-shaft from which the governor is run, and you will probably find it loose. If the engine has been run for any length of time, you will always find the trouble in one of these places; but if it is a new one, the governor-valve might fit a little tight in the valve chamber, and you may have to take it out and use a little emery paper to take off the rough projections on the valve. Never use a file on this valve if you can get emery paper; and I would advise you to always have some of it with you; it will often come handy.

Now, if the engine should start off at a lively gait and continue to run still faster, you must stop at once. The trouble this time is surely in the governor. If the belt is all right, examine the jam-nuts on the top of the governor-valve stem. You will probably find that these nuts have worked loose and the rod is working up, which will increase the speed of the engine. If these are all right, you will find that either a pulley or a little cog-wheel is loose. A quick eye will locate the trouble before you have time to stop. If the belt is loose, the governor will lag, while the engine will run away. If the wheel is loose, the governor will most likely stop, and the engine will go on a tear. If the jam-nut has worked loose, the governor will run on, as usual, except that it will increase its speed as the speed of the engine is increased. Now any of these little things may happen, and are likely to. None of them are serious, provided you take my advice and remain near the engine. Now, if you are thirty or forty feet away from the engine and the governor-belt slips or gets unlaced, or the pulley gets off, about the first thing the engine would do would be to jump out of the belt, and by the time you get to it it will be having a mighty lively time all alone. This might happen once and do no harm, and it might happen again and do a great deal of damage; and you are being paid to run the engine and you must stay by it. The governor is not a difficult thing to handle, but it requires watching.

Now, if I should drop the governor, you might say that I had not given you any instructions about how to regulate it as to speed. I really do not know whether it is worth while to say much about it, for governors are of different design and are necessarily differently arranged for regulating, but to help young learners I will take the Waters governor, which I think the most generally used on threshing and farm engines. By looking at figure 6 you will see two balls, each mounted on one end of a hinged lever; the other end of the lever is attached to a valve-stem, which operates the throttle-valve. If the balls are raised, the valve-stem will drop and shut off the valve; if the balls fall, the valve will be opened wider and more steam admitted to the cylinder. The levers work against a spring after the balls have been raised part way, and this spring tries to keep the balls from rising further. As the engine speeds up, the balls, under the action of centrifugal force, try to rise up, and they do rise a little, and thus shut off the throttle-valve a little, which will make the engine slow down. By changing the tension of the spring you can change the speed at which the engine will run. For doing this you will find on the upper end of the valve- or governor-stem two little brass nuts. The upper one is a thumb-nut, and is made fast to the stem; the

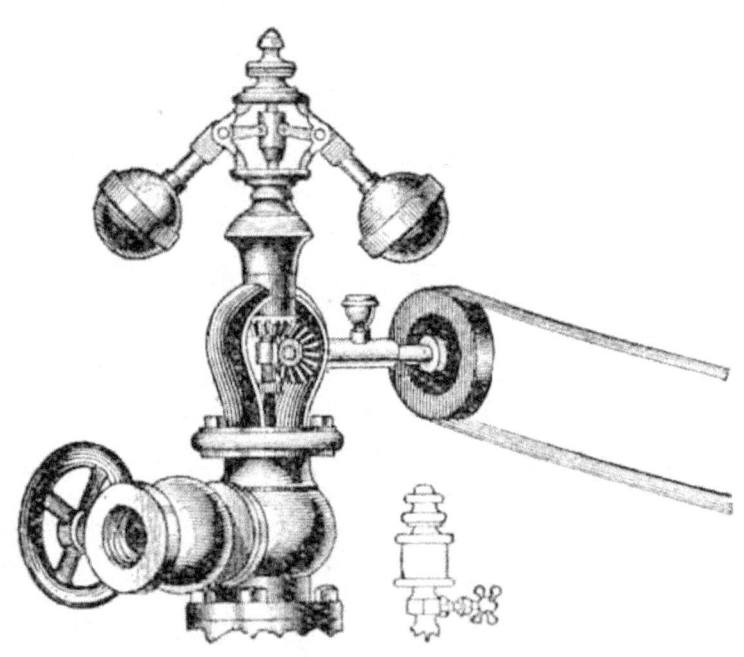

Fig. 6.

second nut is a loose jam-nut. It increases the speed of the engine. Loosening this jam-nut and taking hold of the thumb-nut, you turn it back slowly, watching the motion of your engine all the while; when you have obtained the speed you require, run the thumb-nut down as tight as you can with your fingers; never use a wrench on these nuts. To slow or slacken this speed, loosen the jam-nut as before, except that you must run it up a few turns; then, taking hold of the thumb-nut, turn down slowly until you have the speed required, when you again set the thumb-nut secure. In regulating the speed, be careful not to press down on the stem when turning, as this will make the engine run a little slower than it will after the pressure of your hand is removed.

If, at any time, your engine refuses to start with an open throttle, notice your governor-stem and you will find that it has been screwed down as far as it will go. This frequently happens with a new engine, the stem having been screwed down for its protection in transportation.

In traveling through timber with an engine, be very careful not to let any over-hanging limbs come in contact with the governor.

Now, I think what I have said regarding this particular governor will enable you to handle any one you may come in contact with, as they are all very much alike in these respects. If you will follow the instructions I have given you, the governor will attend to the rest.

PART THIRD.

WATER-SUPPLY.

If you want to be a successful engineer, it is necessary to know all about the pump. I have no doubt that many who read this book can not tell why the old wooden pump (from which he has pumped water ever since he was tall enough to reach the handle) will pump water simply because he works the handle up and down. If you do n't know this, I have quite a task on my hands, for you must not attempt to run an engine until you know the principle of the pump. If you do understand the old town pump, I will not have much trouble with you, for while there is no old-style wooden pump used on the engine, the same principles are used in the cross-head pump. Do not imagine that a cross-head pump means something to be dreaded. It is only a simple lift and force pump, driven from the crosshead. That is where it gets its name, and it does n't mean that you are to get cross at it if it does n't work, for nine times out of ten the fault will be yours.

Water-supply.

Now, I am well aware that all engines do not have cross-head pumps, and with all respect to the builders of engines who do not use them, I am inclined to think that all standard farm engines ought to have a cross-head pump, because it is the most simple pump in use, and is the most economical; and if properly constructed, it is the most reliable. The general arrangement is shown in figure 4.

A cross-head pump consists of a pump barrel, a plunger, one vertical check-valve, and two horizontal check-valves, a globe valve, and one stop-cock, with more or less piping. We will now locate each of these parts, and will then note the part that each performs in process of feeding the boiler.

You will find all or most pump barrels located under the cylinder of the engine. It is placed here for several reasons; it is out of the way; it is a convenient place from which to connect it to the crosshead by which it is driven. On some engines it is located on the top or at the side of the cylinder, and will work equally well. The plunger is connected with the crosshead, and in direct line with the pump barrel, and plays back and forth in the barrel. The vertical check-valve is placed between the pump and the water-supply. It is not absolutely necessary that the first check be a vertical one, but a check of some kind must be so placed. As the water is lifted up to the boiler, it is more convenient to use a vertical check at this point. Just ahead and a few inches from the pump barrel is a horizontal check-valve. Following the course of the water toward the point where it enters the boiler, you will find another check-valve. This is called a "hot-water check." Just below this check, or between it and where the water enters the boiler, you will find a stop-cock. This may be a globe valve; they both answer the same purpose. I will tell you further on why a stop-cock is preferable to a globe valve. While the cross-head pumps may differ as to location and arrangement, you will find that they all require the parts described, and that the checks are so placed that they bear the same relation to each other. No fewer parts can be used in a pump required to lift water and force it against steam pressure. More check-valves may be used, but it would not do to use less. Each has its work to do, and the failure of one defeats all the others. The pump barrel is a hollow cylinder, the chamber being large enough to admit the plunger, which varies in size from $\frac{5}{8}$ of an inch to one inch in diameter, depending upon the size of the boiler to be supplied. The barrel is usually a few inches longer than the stroke of the engine, and is provided at the cross-head end with a stuffing-box and nut. At the discharge end it is tapped out to admit of piping to conduct water from the pump. At the same end, and at the extreme

Water-supply.

end of the travel of the plunger, it is tapped for a second pipe through which the water reaches the pump barrel. The plunger is usually made of steel, and turned down to fit snug in the chamber, and is long enough to play the full stroke of engine between the stuffing-box and point of supply, and to connect with the driver on the crosshead. Now, we will take it for granted that, to begin with, the pump is in good order, and we will start it up stroke at a time and watch its work. We will suppose that we have good water, and a good hard-rubber suction hose attached to the supply pipe just under the globe valve. When we start the pump we must open the little pet cock between the two horizontal check-valves. The globe valve must be open so as to let the water in. A check-valve, whether it is vertical or horizontal, will allow water to pass through it one way only, if it is in good working order. If the water will pass through both ways, it is of no account. Now, the engine starts out on the upward stroke and draws the plunger out of the chamber. This leaves a space in the barrel which must be filled. Air can not get into it, because the pump is in perfect order; neither can the air get to it through the hose, as it is in the water, so that the pressure on the outside of the water causes it to flow up through the pipes, through the first check-valve and into the pump barrel, and fills the space; and if the engine has a twelve-inch stroke, and the plunger is one inch in diameter, we have a column of water in the pump twelve inches long and one inch in diameter.

The engine has now reached its outward stroke and starts back. The plunger comes back with it and takes the space occupied by the water, which must get out of the way for the plunger. The water came up through the first check-valve, but it can't get back that way. There is another check-valve just ahead, and as the plunger travels back it drives the water through this second check. When the plunger reaches the end of the backward stroke it has driven the water all out. It then starts forward again, but the water which has been driven through the second check can't get back and the plunger continues to force more water through the second check, taking four or five strokes of the plunger to fill the pipes between the second check-valve and the hot-water check-valve. If the gauge shows 100 pounds steam, the hot-water check is held shut by 100 pounds pressure, and, when the space between the check-valves is filled with water, the next stroke of the plunger will force the water through the hot-water check-valve, and the valve is held shut by the 100 pounds steam pressure so that the pump must force the water against this hot-water check-valve with a force greater than 100 pounds pressure. If the pump is in good condition, the plunger does its work and the water is forced through into the boiler.

A clear, sharp click of the valves at each stroke of the plunger is certain evidence that the pump is working well.

The small drain cock between the horizontal checks is placed there to assist in starting the pump, to tell when the pump is working, and to drain the water off to prevent freezing. When the pump is started to work and this drain cock is opened, and the hot water in the pipes drained off, the globe valve is then opened, and, after a few strokes of the plunger, the water will begin to flow out through the drain cock, which is then closed, and you may be reasonably certain that the pump is working all right. If at any time you are in doubt as to whether the pump is forcing the water through the pipes, you can easily ascertain by opening this drain cock. It will always discharge cold water when the pump is working. Another way to tell if the pump is working is by placing your hand on the first two check-valves. If they are cold, the pump is working all right, but if they are warm, the cold water is not being forced through them.

It is very important when the pump does n't work to ascertain what the trouble is. If it should stop suddenly, examine the tank and ascertain if you have any water. If you have sufficient water, it may be that there is air in the pump chamber, and the only way that it can get in is through the stuffing-box around the plunger, if the pipes are all tight. Give this stuffing nut a turn, and if the pump starts off all right you have found the trouble, and it would be well to repack the pump the first chance you get.

If the trouble is not in the stuffing-box, go to the tank and see if there is anything over the screen or strainer at the end of the hose. If there is not, take hold of the hose and you can tell if there is any suction. Then ascertain if the water flows in and then out of the hose again. You can tell this by holding your hand over the end of the hose. If you find that it draws the water in and then forces it out again, the trouble is with the first check-valve. There is something under it which prevents its shutting down. If, however, you find that the water is not forced out of the hose and returned to the tank, examine the second check. If there were something under it, it would prevent the pump working, because the pump forces the water through it; and, as the plunger starts back, if the check fails to hold, the water flows back and fills the pump barrel again.

The trouble may, however, be in the hot-water check, and it can always be told whether it is in the second check or hot-water check by opening the little drain cock. If the water which goes out through it is cold, the trouble is in the second check; but if hot water and steam are blown out through this little drain cock, the trouble is in the hot-water check, or the one next to the boiler. This check must never be tampered with without first turning the stop-cock between this check and the boiler. The valve can then be taken out and the obstruction removed. Be very careful never to take out the hot-water check without closing the stop-cock, for if you do you will get badly scalded; and never start the pump without opening this valve, for if you do it will burst the pump.

The obstruction under the valves is sometimes hard to find. A young man in southern Iowa got badly fooled by a little pebble about the size of a pea, which got into the pipe, and when he started his pump the pebble would be forced up under the check and let the water back. When he took the check out the pebble was not there, for it had dropped back into the pipe. You will see that it is necessary to make a careful examination, and not get mad, pick up a wrench, and whack away at the check-valve, bruising it so that it will not work. Remember that it would work if it could, and make up your mind to find out why it does n't work. A few years ago I was called several miles to see an engine on which the pump would not work. The engine had been idle for two years and the engineer had been trying all that time to make the pump work. I took the cap off of the horizontal check, just forward of the pump barrel, and took the valve out and discovered that the check was reversed. I told the engineer that if he would put the check in so that the water could get through he would have no more trouble. This fellow had lost his head; he was completely rattled; he insisted that the valve had always been on that way, although the engine had been run two years.

There are other causes that would prevent the pump working besides lack of packing and obstructions under the valves: the valve may stick; when it is raised to allow the water to flow through, it may stick in the valve chamber and refuse to settle back in the seat. This may be caused by a little rough place in the chamber, or a little projection on the valve, and can generally be remedied by tapping the under side of check with a wrench or hammer. Do not strike it so hard as to bruise the check, but simply tap it as you would tap an eggshell without breaking it. If this does n't remedy the trouble, take the valve out, bore a hole in a board about $\frac{1}{2}$ of an inch deep and large enough to permit the valve to be turned. Drop a little emery dust in this hole. If you have n't any emery dust, scrape some grit from a common whetstone. If you have no whetstone, put some fine sand or gritty

Water-supply.

soil in the hole, put the valve on top of it, put your brace on the valve, and turn it vigorously for a few minutes, and you will remove all roughness.

Constant use may sometimes make a burr on the valve which will cause it to stick. Put it through the above course and it will be as good as new. If this little process were generally known, a great deal of trouble and annoyance could be avoided.

It will not be necessary to describe other styles of pumps. If you know how to run the cross-head pump, you can run any of the others. Some engines have a cross-head pump only; others have an independent pump; others have an injector, or inspirator, and some have both cross-head pump and injector. I think a farm engine should be supplied with both.

It is neither wise nor necessary to go into a detailed description of an injector. The young reader will be likely to become convinced if an injector works for five minutes it will continue to work, if the conditions remain the same. If the water in the tank does not become heated, and no foreign substance is permitted to enter the injector, there is nothing to prevent its working properly. An injector will not pump hot water, neither will it start to work while it is hot. In an injector the size that is usually used on farm engines, the opening through which the water passes is not over $\frac{1}{16}$ of an inch in diameter.

It will be readily seen that muddy water can not be used.

The injector should not be placed too near the boiler, as the heat from the boiler will make it hard to start the injector each time after it has been standing idle.

If the injector is so hot that it will not lift the cold water, there is no way of cooling it except by applying the water on the outside. This is most effectively done by covering the injector with a cloth and pouring water over the cloth. If after the injector has become cool it still refuses to work, you may be sure that there is some obstruction in it that must be removed. This can be done by taking off the cap, or plug-nut, and running a fine wire through the cone valve and cylinder valve.

The above suggestions with reference to an injector refer more especially to the Eberman injector, and others of its class that require only the steam globe to be manipulated in order to start the injector to work. There are other makes that require, first, that the steam supply be opened; then open the globe, which permits the water to reach the injector. If you have an injector of this kind, when you come to start it you must first give it a sufficient head of steam, then open the globe valve, and when the water is lifted and begins to discharge from the overflow, let it run, say, about ten seconds, then shut the water off for about one second, and then open up the globe with a quick turn, and the injector will start to work without trouble. If this style of injector will not work, it can be remedied the same as the Eberman and others of its class. It must be borne in mind that the injector must be fitted perfectly air-tight, and that the steam supply must be tight.

To start and work the Penberthy injector these directions will be useful :

To start at twenty-five or thirty pounds of steam pressure, open water valve one turn and open steam valve full, when it should work all right, water going into the boiler. If water comes out the overflow, shut off valves and try it again, opening the water valve a little less. With steam at eighty pounds it is not necessary to juggle the water valve ; just open it wide and then turn on steam valve full.

It is desirable to feed the water hot into the boiler. To do this, adjust the water valve until by feeling the pipe leading to the boiler you find that it is getting warm.

It is now time to give some attention to the heater. While the heater is no part of the pump, it is connected with it and does its work between the two horizontal check-valves. Its purpose is to heat the water before it passes into the boiler. The water on its way from the pump to the boiler is forced through a coil of pipes around which the exhaust steam passes on its way from the cylinder to the exhaust nozzle in the smoke-stack.

The heaters are made in several different designs, but it is not necessary to describe all of them, as they require little attention, and they all answer the same purpose. The most of them are made by the use of a hollow bed-plate with steam-fitted heads or plates. The water pipe passes through the plate at the end of the heater into the hollow chamber, and a coil of pipes is formed, and the pipe then passes back through the head or plate to the hot-water check-valve and into the boiler.

The steam enters the cylinder from the boiler, varying in degrees of heat from 400 to 600. After acting on the piston head, it is exhausted directly into the chamber or hollow bed-plate through which the pipes pass. The water, when it enters the heater, is as cold as when it left the tank, but the steam which surrounds the pipes has lost but little of its heat, and by the time the water passes through the coil of pipes it is heated to nearly boiling point and can be introduced into the boiler with little tendency to reduce the steam. This use of the exhaust steam is economical, as it saves fuel, and it will be injurious to pump cold water direct into a hot boiler.

If your engine is fitted with both cross-head pump and injector, you use the injector for pumping water when the engine is not running. The injector heats the water almost as hot as the heater. If your engine is running and doing no work, use your injector and stop the pump,

Water-supply.

for, while the engine is running light, the small amount of exhaust steam is not sufficient to heat the water and the steam will be reduced rapidly. You will understand, therefore, that the injector is intended principally for an emergency rather than for general use. It should always be kept in order, for should the pump decline to work, you have only to start your injector and use it until such time as you can remedy the trouble.

I said a little while ago that I thought that every traction engine ought to be supplied with both a pump and an injector. Before I close the subject I wish to call your attention to the Clark pump generally supplied with the North West Traction Engine. Fig. 17 gives you a view of this engine pump. It is an independent steam-pump of a vertical type, generally located on the boiler close to the traction gear and on the side opposite to the steering wheel handle.

The engine and the pump are permanently connected together, and have their cylinder located one above the other and on the same line. The engine is supplied with its own cylinder, steam-chest, valves, etc., and the pump is also complete in itself.

This pump differs, therefore, from the cross-head pump in that it is absolutely independent of the operation of the engine and that it can be used when the engine is not running; and, further, that it can be run at any speed wanted independent of the speed of the engine, thus enabling the water-supply to be regulated to suit all conditions.

The exhaust steam from the pump is utilized for heating the feed-water before it is pumped into the boiler, which results in the advantage that the water supplied to the boiler is of the same temperature whether the engine is running or not running. The advantage of being able to supply the boiler with feed-water of the same temperature under all conditions is very great, as it materially prolongs the life of the boiler. Another advantage secured by using an independent pump is that it can also be used by hand, which enables you to supply your

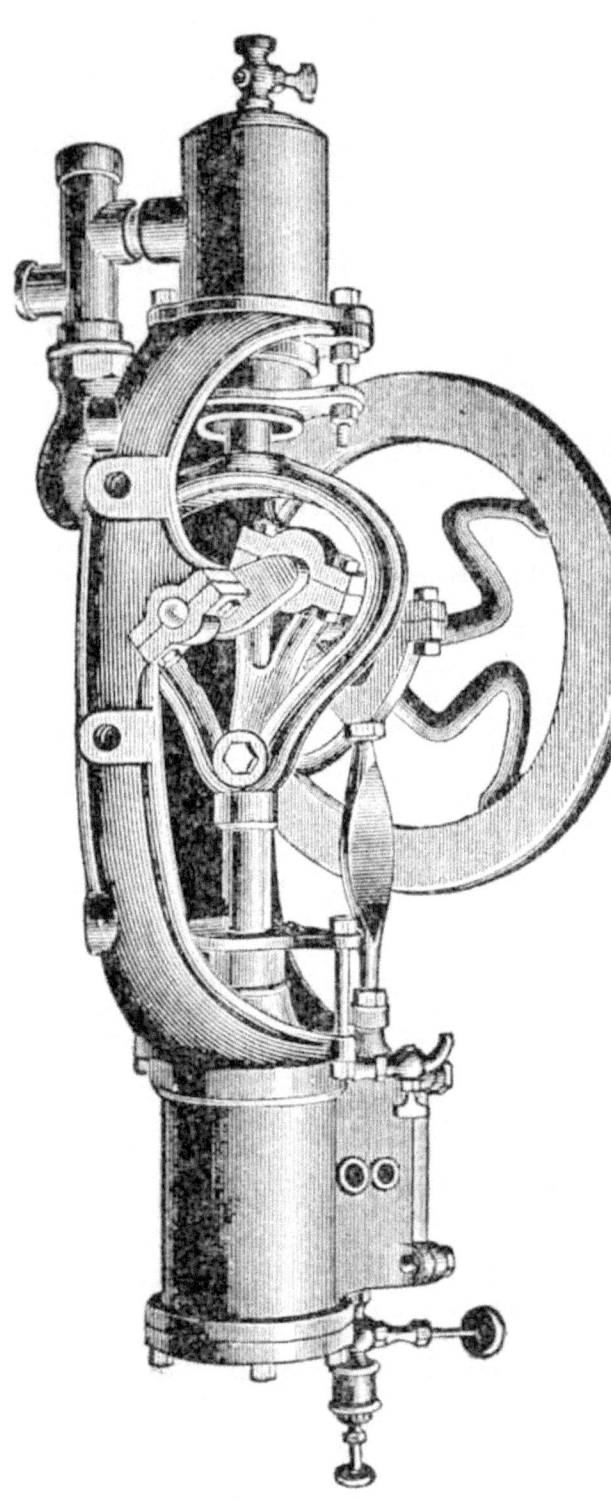

FIG. 17.

Water-supply.

boiler with water even if you should not have any steam at all.

The pump itself is extremely simple in construction and can be regulated for any speed which you may desire. If it should be necessary to fill your boiler rapidly, you can do so, and afterward you can reduce the speed to such a number of revolutions that the pump will only replace the amount of water which has been evaporated.

We have now explained how you get your water-supply. You understand that you must have water first and then fire. Be sure that you have the water-supply first.

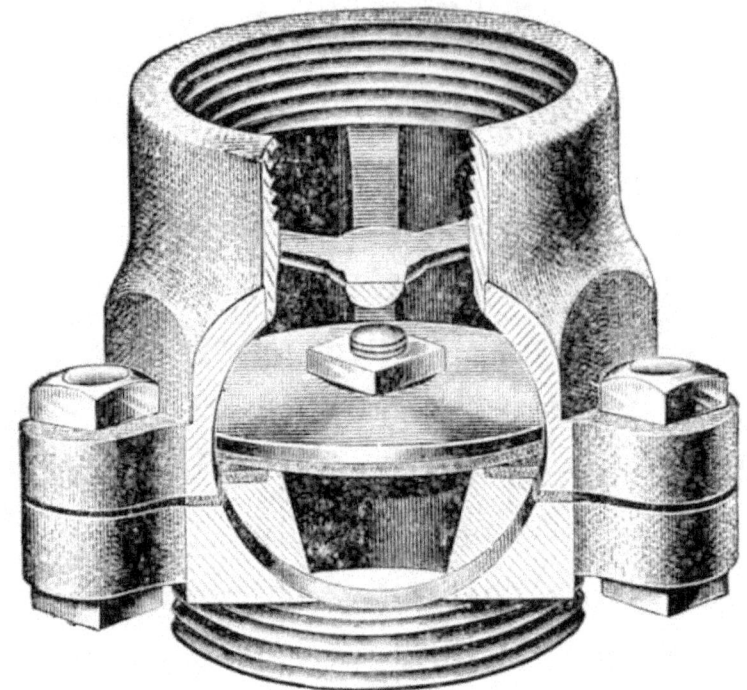

Fig. 11.—Check Valve.

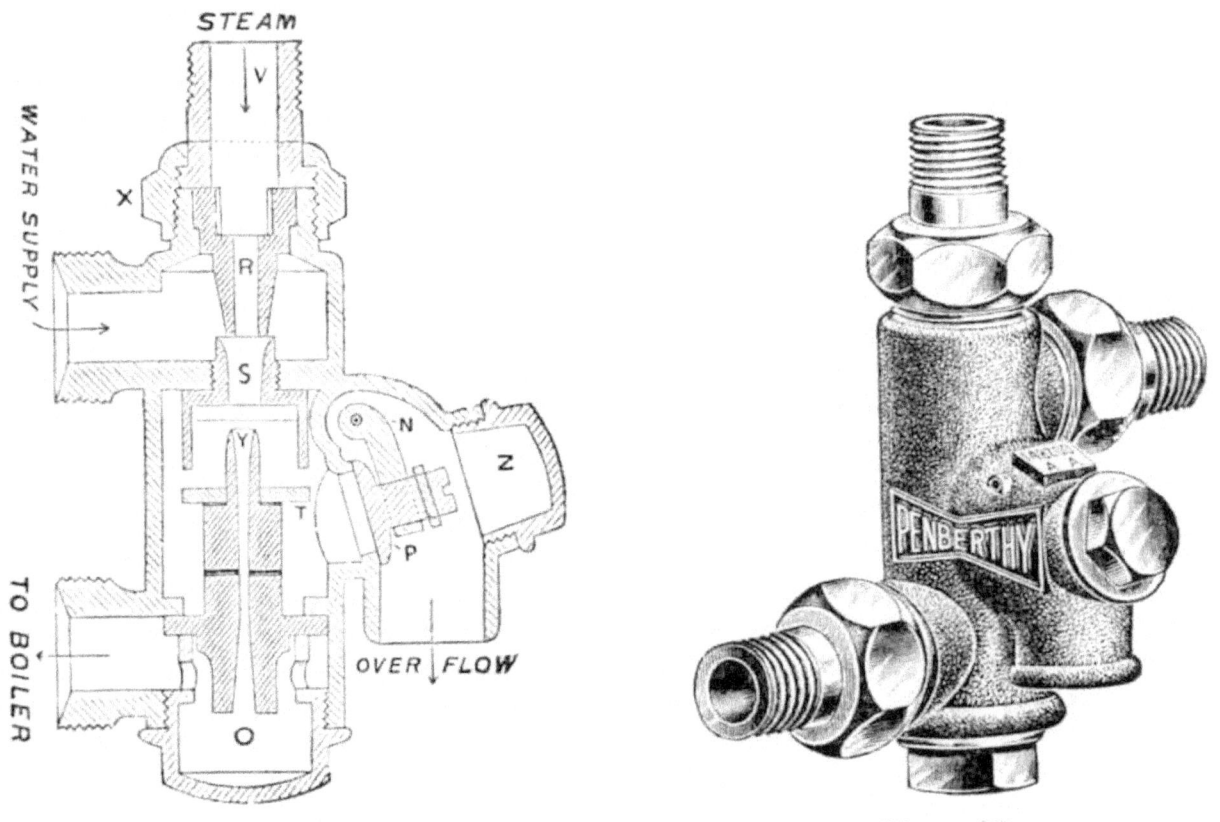

Fig. 12. Fig. 13.

Penberthy Injector.

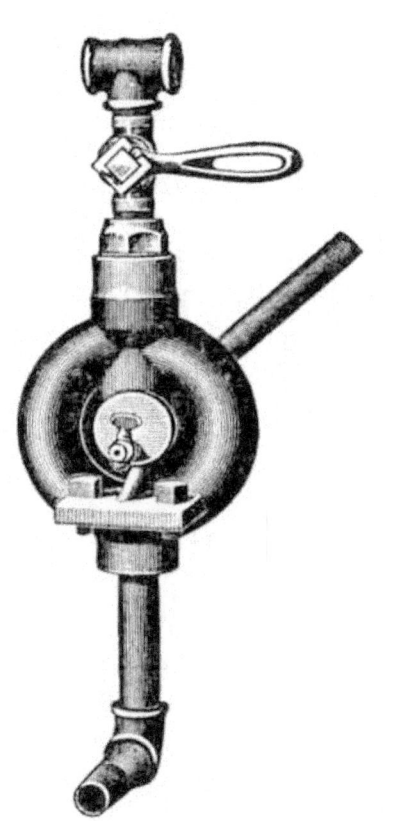

Fig. 16.—The Siphon.

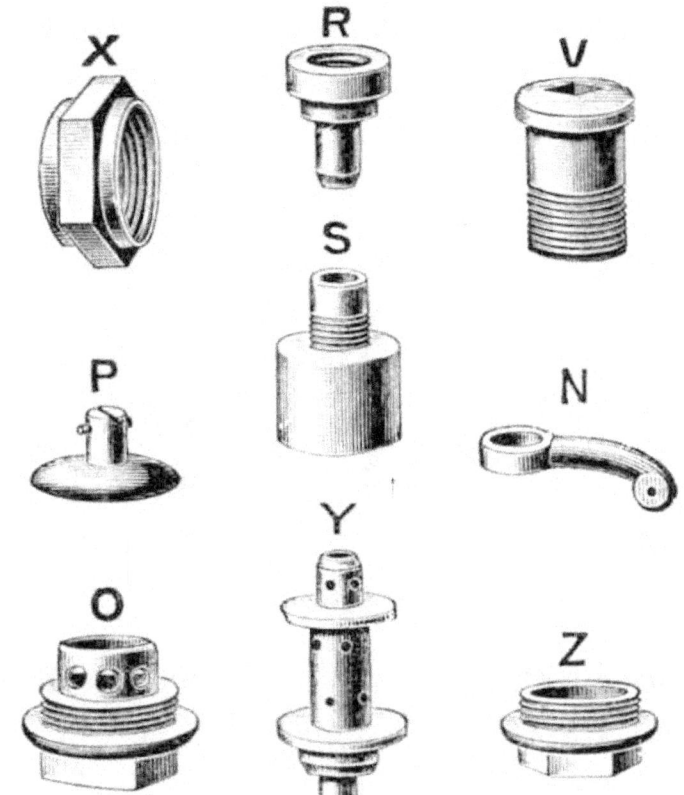

R—STEAM JET. ⋯⋯⋯⋯V—TAIL PIPE.⋯⋯
S—SUCTION JET. ⋯⋯⋯⋯X—COUPLING NUT.
Y—DELIVERY JET.⋯⋯⋯⋯N—OVERFLOW HINGE.
O—PLUG. ⋯⋯⋯⋯⋯⋯⋯⋯P—OVERFLOW VALVE
⋯⋯Z—OVERFLOW CAP⋯⋯

Fig. 14.—Details of a Penberthy Injector.

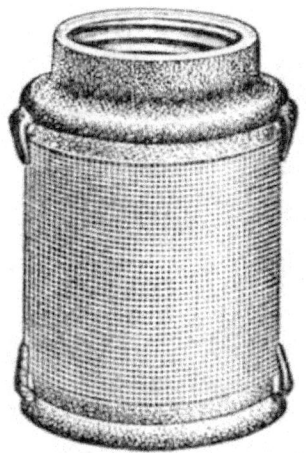

Fig. 15.—Strainer for Hose Connection.

PART FOURTH.

THE BOILER.

A boiler should be kept clean outside and inside—outside for your own credit and inside for the credit of the manufacturers. A dirty boiler requires hard firing, takes lots of fuel, and is unsatisfactory in every way.

The best way to keep it clean is not to let it get dirty. The place to begin work is with your " water boy ; " persuade him to be very careful of the water he brings you ; if you can't succeed in this, ask him to resign.

I have seen a water hauler back into a stream and then dip the water from the lower side of the tank ; the muddy water always goes down stream and the wheels stir up the mud, and your bright water hauler dips it into the tank. While if he had dipped it from the upper side, he would have got clear water. However, the days of dipping water are past, but a water boy that will do as I have stated is just as liable to throw his hose into the muddy water or lower side of tank as on the upper side, where it is clear.

The Boiler.

See that he keeps his tank clean. We have seen tanks with one-half inch of mud in the bottom. We know that there are times when you are compelled to use muddy water, but as soon as it is possible to get clear water make him wash out his tank and don't let him haul it around till the boiler gets it all.

Allow me just here to tell you how to construct a good tank for a traction engine. You can make the dimensions to suit yourself, but across the front end, and about two feet back, fit a partition or second head; in the center of this head and about an inch from the bottom bore a two-inch hole. Place a screen over this hole on the side next the rear, and on the other side, or side next front end, put a valve. You can construct the valve in this way: Take a piece of thick leather, about four inches long, and $2\frac{1}{2}$ inches wide; fit a block of wood (a large bung answers the purpose nicely) on one end, trimming the leather around one side of the wood; then nail the long part of the valve just above the hole so that the valve will fit nicely over the hole in partition. When properly constructed, this valve will allow the water to flow into the front end of the tank, but will prevent its running back. So, when you are on the road with part of a tank of water, and start down hill, this front part fills full of water, and when you start up hill it can not get back, and your pumps will work as well as if you had a full tank of water. As most all tanks are tapered at their lower front end, you can not get your pumps to work well in going up a steep hill with anything less than a full tank. Now, this may be considered a little out of the engineer's duty, but it will save lots of annoyance if he has his tank supplied with this little appliance, which is simple but does the business.

A boiler should be washed out and not blown out. I believe I am safe in saying that more than half the engineers of threshing engines to-day depend on the "blowing out" process to clean their boilers. I do n't intend to tell you to do anything without giving my reasons. We will take a hot boiler, for instance, say fifty pounds steam. We will, of course, take out the fire. It is not supposed that any one will attempt to blow out the water with any fire in the fire-box. We will, after removing the fire, open the blow-off valve, which will be found at the bottom or lowest water point. The water is forced out very rapidly with this pressure, and the last thing that comes out is the steam. This steam keeps the entire boiler hot till everything is blown out, and the result is that all the dirt, sediment, and lime is baked solid in the tubes and side of fire-box. But you say you know enough to not blow off at fifty pounds pressure. Well, we will say five pounds then. You will admit that the boiler is not cold by any means, even at only five pounds, and if you know enough not to blow off at fifty pounds, you certainly know that at

The Boiler.

five pounds pressure the damage is not entirely avoided. As long as the iron is hot the dirt will dry out quickly, and by the time the boiler is cold enough to force cold water through it safely, the mud is dry and adheres closely to the iron. Some of the foreign matter will be blown out, but you will find it a difficult matter to wash out what sticks to the hot iron.

I am aware that some engineers claim that the boiler should be blown out at about five pounds or ten pounds pressure, but I believe in taking the common sense view. They will advise you to blow out at a low pressure, and then, as soon as the boiler is cool enough, to wash it thoroughly.

Now, if you must wait till the boiler is cool before washing, why not let it cool with the water in it? Then, when you let the water out, your work is easy, and the moment you begin to force water through it, you will see the dirty water flowing out at the man or hand hole. The dirt is soft and washes very easily; but if it had dried on the inside of the boiler while you were waiting for it to cool, you would find it very difficult to wash off.

You say I said to force the water through the boiler, and to do this you must use a force pump. No engineer ought to attempt to run an engine without a force pump. It is one of the necessities. You say, can't you wash out a boiler without a force pump? Oh, yes! You can do it just like some people do business. But I started out to tell you how to keep your boiler clean, and the way to do it is to wash it out, and the way to wash it out is with a good force pump. There are a number of good pumps made, especially for threshing engines. They are fitted to the tank for lifting water for filling, and are fitted with a discharge hose and nozzle.

You will find at the bottom of boiler one or two hand-hole plates,—if your boiler has a water bottom,—if not, they will be found at the bottom of sides of fire-box. Take out these hand-hole plates. You will also find another plate near the top, on fire-box end of boiler; take this out, then open up smoke-box door and you will find another hand-hole plate near the lower row of tubes; take this out, and then you are ready for your waterworks, and you want to use them vigorously; don't throw in a few buckets of water, but continue to direct the nozzle to every part of the boiler, and don't stop as long as there is any muddy water flowing at the bottom hand holes. This is the way to clean your boilers, and don't think that you can be a success as an engineer without this process, and once a week is none too often. If you want satisfactory results from your engine you must keep a clean boiler, and to keep it clean requires care and labor. If you neglect it you can expect trouble. If you blow out your boiler hot, or if the mud and slush bakes on the tubes, there is soon a scale formed on the tubes, which decreases the boiler's evaporating capacity. You, therefore, in order to make sufficient amount of steam, must increase the amount of fuel, which of itself is a source of expense, to say nothing of extra labor and the danger of causing the tubes to leak from the increased heat you must produce in the fire-box in order to make steam sufficient to do the work.

You must not expect economy of fuel, and keep a dirty boiler, and do n't condemn a boiler because of hard firing until you know it is clean; and do n't say it is clean when it can be shown to be half full of mud.

SCALE.

Advertisements say that certain compounds will prevent scale on boilers, and I guess they tell the truth, as far as they go; but they do n't say what the result may be on iron. I will not advise the use of any of these preparations, for several reasons. In the first place, certain chemicals will successfully remove the scale formed by water charged with bicarbonate of lime, and have no effect on water charged with sulphate of lime. Some kinds of bark—sumac, logwood, etc.—are sufficient to remove the scale from water charged with magnesia or carbonate of lime, but they are injurious to the iron owing to the tannic acid with which they are charged. Vinegar, rotten apples, slop, etc., owing to their containing acetic acid, will remove scale, but this is even more injurious to the iron than the barks. Alkalies of any kind, such as soda, will be found good in water containing sulphate of lime, by converting it into a carbonate and thereby forming a soft scale, which is easily washed out; but these have their objections, for, when used to excess, they cause foaming. Most commercial compounds are, however, nothing but strong alkalies.

Petroleum is not a bad thing in water where sulphate of lime prevails; but you should use only the refined, as crude oil sometimes helps to form a very injurious scale. Moreover, if used to excess it gives a great deal of trouble by leaking out through the joints of the boiler tubes.

CLEAN FLUES.

We have been urging you to keep your boiler clean. Now, to get the best results from your fuel it will also be necessary to keep your flues clean; as soot and ashes are non-conductors of heat, you will find it very difficult to get up steam with a coating of soot in your tubes. Most factories furnish with each engine a flue cleaner and rod. This cleaner should be made to fit the tubes snugly, and should be forced through each separate tube every morning before building a fire. Some engineers never touch their flues with a cleaner, but when they choke the ex-

The Boiler.

haust sufficiently to create such a draft as to clean the flues, they are working the engine at a great disadvantage, besides being much more liable to pull the fire out at the top of smoke-stack. If it were not necessary to create draft by reducing your exhaust nozzle, your engine would run much nicer and be much more powerful if your nozzle was not reduced at all. However, you must reduce it sufficiently to give draft, but don't impair the power by making the engine clean its own flues. I think ninety per cent. of the fires started by traction engines can be traced to the engineer having his engine choked at the exhaust nozzle. This is dangerous for the reason that the excessive draft created throws fire out at the stack. It cuts the power of the engine by creating back pressure. We will illustrate this: Suppose you close the exhaust entirely, and the engine would not turn itself. If this is true, you can readily understand that partly closing it will weaken it to a certain extent. So, remember that the nozzle has something to do with the power of the engine, and you can see why the fellow that makes his engine clean its own flues is not the brightest engineer in the world.

While it is not my intention to encourage the foolish habit of pulling engines, to see which is the best puller, however, should you get into this kind of a test, you will show the other fellow a trick by dropping the exhaust nozzle off entirely, and no one need know it. Your engine will not appear to be making any effort, either, in making the pull. Many a test has been won more through the shrewdness of the operator than the superiority of the engine.

The knowing of this little trick may also help you out of a bad hole some time when you want a little extra power. And this brings us to the point to which I want you to pay special attention. The majority of engineers, when they want a little extra power, give the safety-valve a twist.

Now, I have already told you to carry a good head of steam. Anywhere from 100 to 120 pounds of steam is good pressure for a threshing engine, and 115 pounds is a nice pressure, and is plenty; and if you have your valve set to blow off at 115, let it be there, and don't screw it down every time you want more power, for if you do you will soon have it up to 125; and should you want more steam at some other time, you will find yourself screwing it down again, and what was really intended for a safety-valve loses all its virtue as a safety, as far as you and those around you are concerned. If you know you have a good boiler, you are safe in setting it at 125 pounds, provided you are determined to not set it up to any higher pressure. But my advice to you is that if your engine will not do the work required of it at 115 pounds, you had

The Boiler.

best do what you can with it until you can get a larger one.

A safety-valve is exactly what its name implies, and there should be a heavy penalty for any one taking that power away from it.

If you refuse to set your safety down at any time, it does not imply that you are afraid of your boiler, but, rather, that you understand your business and realize your responsibility.

I stated before what you should do with the safety-valve in starting a new engine. You should also attend to this part of it every few days. See that it does not become slow to work. You should note the pressure every time it blows off; you know where it ought to blow off, so do n't allow it to stick or hold the steam beyond this pressure. If you are careful about this, there is no danger about it sticking some time when you do n't happen to be watching the gauge. The steam-gauge will tell you when the pop ought to blow off, and you want to see that it does it.

STEAM-GAUGE.

Some engineers call a steam-gauge a "clock." I suppose they do this because they think it tells them when it is time to throw in coal, and when it is time to quit, and when it is time for the safety-valve to blow off. If that is what they think a steam-gauge is for, I can tell them that it is time for them to learn differently.

It is true that in a certain sense it does tell the engineer when to do certain things, but not as a clock would tell the time of day. The office of a steam-gauge is to enable you to read the pressure on your boiler at all times, the same as a scale will enable you to determine the weight of any object.

As this is the duty of the steam-gauge, it is necessary that it be absolutely correct. By the use of an unreliable gauge you may become thoroughly bewildered, and, in reality, know nothing of what pressure you are carrying.

This will occur in about this way: Your steam-gauge becomes weak, and if your safety is set at 100 pounds, it will show 110, or even more, before the pop allows the steam to escape; or, if the gauge becomes clogged, the pop may blow off when the gauge only shows ninety pounds or less. This latter is really more dangerous than the former, as you would most naturally conclude that your safety was getting weak, and about the first thing you would do would be to screw it down so that the gauge would show 100 before the pop would blow off, when, in fact, you would have 110 or more.

So you can see at once how important it is that your gauge and safety should work exactly together, and there is but one way to make certain of this, and that is to test

The Boiler.

your steam-gauge. If you know the steam-gauge is correct, you can make your safety-valve agree with it; but never try to make it do it till you know the gauge is reliable.

HOW TO TEST A STEAM-GAUGE.

Take it off and take it to some shop where there is a steam boiler in active use; have the engineer attach your gauge where it will receive the direct pressure, and if it shows the same as his gauge, it is reasonable to suppose that your gauge is correct. If the engineer to whom you take your gauge should say he thinks his gauge is weak, or a little strong, then go somewhere else. I have already told you that I did not want you to think anything about your engine—I want you to know it. However, should you find that your gauge shows, when tested with another gauge, that it is weak or unreliable in any way, you want to repair it at once, and the safest way is to get a new one; and yet I would advise you first to examine it and see if you can not discover the trouble. It frequently happens that the pointer becomes loosened on the journal or spindle which attaches it to the mechanism that operates it. If this is the trouble, it is easily remedied, but should the trouble prove to be in the spring, or the delicate mechanism, it would be much more satisfactory to get a new one.

In selecting a new gauge you will be better satisfied

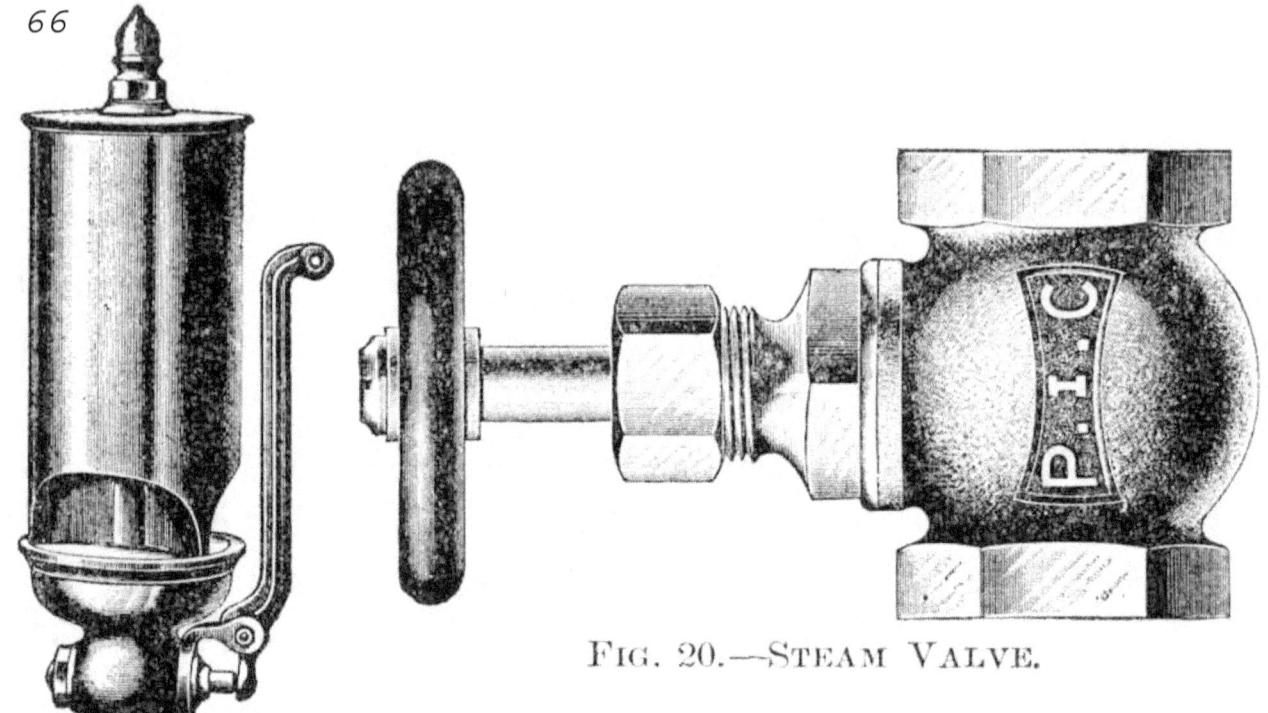

Fig. 18.—Whistle.

Fig. 20.—Steam Valve.

Fig. 19.—Steam Gauge.

Fig. 21.—Safety Valve.

with a gauge having a double spring or tube, as they are less liable to freeze or become strained from a high pressure, and the double spring will not allow the needle or pointer to vibrate when subject to a shock or sudden increase of pressure, as with the single spring. A careful engineer will have nothing to do with a defective steam-gauge or an unreliable safety-valve. Some steam-gauges are provided with a seal, and as long as this seal is not broken, the factory will make it good.

FUSIBLE PLUG.

We have told you about a safety-valve; we will now have something to say of a safety plug. A safety, or fusible, plug is a hollow brass plug or bolt, screwed into the top of the crown sheet, the hole through the plug being filled with some soft metal that will fuse at a much less temperature than is required to burn iron. The heat from the fire-box will have no effect on this fusible plug as long as the crown sheet is covered with water, but the moment that the water level falls below the top of the crown sheet, thereby exposing the plug, this soft metal is melted and runs out, allowing the steam to rush down through the opening in the plug, putting out the fire and preventing any injury to the boiler. This all sounds very nice, but I am free to confess that I am not an advocate of a fusible plug. After telling you to never allow the water to get

The Boiler.

low, and then to say there is something to even make this allowable, sounds very much like the preacher who told his boy "never to go fishing on Sunday, but if he did go to be sure and bring home the fish." I would have no objection to the safety plug if the engineer did not know it was there. I am aware that some States require that all engines be fitted with a fusible plug. I do not question their good intentions, but I do question their good judgment. It seems to me they are granting a license to carelessness. For instance, an engineer is running with a low gauge of water, owing possibly to the tank being delayed longer than usual; he knows the water is getting low, but he says to himself, "well, if the water gets too low I will only blow out the plug," and so he continues to run until the tank arrives. If the plug holds, he at once begins to pump in cold water, and most likely does it on a very hot sheet, which, of itself, is something he never should do; and if the plug does blow out, he is delayed a couple of hours, at least, before he can put in a new plug and get up steam again. Now, suppose he had not had a soft plug (as they are sometimes called): he would have stopped before he had low water. He would not even have had a hot crown sheet, and would only have lost the time he waited on the tank. This is not a fancied circumstance by any means, for it happens every day. The engineer running an engine with a safety plug seldom stops for a load of water until he blows out the plug. It frequently happens that a fusible plug becomes corroded to such an extent that it will stand a heat sufficient to burn the iron. This is my greatest objection to it. The engineer continues to rely on it for safety, the same as if it were in perfect order, and the ultimate result is he burns or cracks his crown sheet. I have already stated that I have no objection to the plug if the engineer did not know it was there, so if you must use one, attend to it, and every time you clean your boiler scrape the upper or water end of the plug with a knife, and be careful to remove any corrosive matter that may have collected on it; and then treat your boiler exactly as though there were no such a thing as a safety plug in it. A safety plug was not designed to let you run with any lower gauge of water. It is placed there to prevent injury to the boiler in case of an accident, or when, by some means, you might be deceived in your gauge of water, or if, by mistake, a fire was started without any water in the boiler.

Should the plug melt out, it is necessary to replace it at once or as soon as the heat will permit you to do so. It might be a saving of time to have an extra plug always ready, then all you have to do is to remove the melted one by unscrewing it from the crown sheet and screwing the extra one in. But if you have no extra plug you must remove the first one and refill it with babbitt. You can do this by filling one end of the plug with wet clay and pouring the metal into the other end, and then pounding it down smooth to prevent any leaking. This done, you can screw the plug back into its place.

If you should have two plugs, as soon as you have melted out one replace it with the new one and refill the other at your earliest convenience. By the time you have replaced a fusible plug a few times in a hot boiler you will conclude it is better to keep water over your crown sheet.

LEAKY FLUES.

What makes flues leak? I asked this question once, and the answer was that the flues were not large enough to fill up the hole in flue sheet. This struck me as being funny at first, but on second thought I concluded it was about correct. Flues may leak from several causes, but usually it can be traced to the carelessness of some one. You may have noticed before this that I am inclined to blame a great many things on carelessness. Well, by the time you have run an engine a year or two you will conclude that I am not unjust in my suspicions. I do not blame engineers for everything, but I do say that they are responsible for a great many things which they endeavor to shift on to the manufacturer. If the flues in a new boiler leak, it is evident that they were slighted by the boiler-maker; but should they run a season, or part of a season, before leaking, then it would indicate that the boiler-maker did his duty, but the engineer did not do his. He has been building too hot a fire to begin with, or has been letting his fire door stand open; or he may have overtaxed his boiler; or else he has been blowing out his boiler when too hot; or has at some time blown out with some fire in fire-box. Now, any one of these things repeated a few times will make the best of them leak. You have been advised already not to do these things, and if you do them, or any one of them, I want to know what better word there is to express it than "carelessness."

There are other things that will make your flues leak. Pumping cold water into a boiler with a low gauge of water will do it, if it does nothing more serious. Pouring cold water into a hot boiler will do it. For instance, if for any reason you should blow out your boiler while in the field, and as you might be in a hurry to get to work, you would not let the iron cool before beginning to refill. I have seen an engineer pour water into a boiler as soon as the escaping steam would admit it. The flues can not stand such treatment, as they are thinner than the shell or flue sheet, and therefore cool much quicker, and in contracting are drawn from the flue sheet, and, as a matter of course, must leak. A flue, when once started to leak, seldom stops without being set up, and one leaky flue will start others, and what are you going to do about it? Are you

The Boiler.

going to send to a boiler shop and get a boiler-maker to come out and fix them and pay him from forty to sixty cents an hour for doing it? I do n't know but that you must the first time, but if you are going to make a business of making your flues leak, you had best learn how to do it yourself. You can do it if you are not too big to get into the fire door. You should provide yourself with a flue expander and a calking tool with a machinist's hammer (not too heavy). Take into the fire-box with you a piece of clean waste with which you will wipe off the ends of the flues and flue sheet to remove any soot or ashes that may have collected around them. After this is done you will force the expander into the flues, driving it well up in order to bring the shoulder of expander up snug against the head of the flues. Then drive the tapering pin into the expander. By driving the pin in too far you may spread the flue sufficient to crack it, or you are more liable, by expanding too hard, to spread the hole in flue sheet and thereby loosen other flues. You must be careful about this. When you think you have expanded sufficient, hit the pin a side blow in order to loosen it and turn the expander about one-quarter of a turn, and drive it up as before; loosen up and continue to turn as before until you have made the entire circle of flues. Then remove the expander and you are ready for your beader or calking tool. It is best to expand all the flues that are leaking before beginning with the beader.

The beader is used by placing the gauge or guide end within the flue, and with your light hammer the flue can be calked or beaded down against the flue sheet. Be careful to use your hammer lightly so as not to bruise the flues or sheet. When you have gone over all the expanded flues in this way, you (if you have been careful) will not only have a good job, but will conclude that you are somewhat of an expert at it. I never saw a man go into a firebox and stop the leak but that he came out well pleased with himself. The fact that a fire-box is no pleasant workshop may have something to do with it. If your flues have been leaking badly, and you have expanded them, it would be well to test your boiler with cold water pressure to make sure that you have a good job.

How are you going to test your boiler? If you can attach to a hydrant do so, and when you have given your boiler all the pressure you want, you can then examine your flues carefully, and should you find any leaking of water, you can use your beader lightly until all such leaks are stopped. If the waterworks will not afford you sufficient pressure you can bring it up to the required pressure by attaching a hydraulic pump or a good force pump.

In testing for the purpose of ascertaining if you have

The Boiler.

a good job on your flues, it is not necessary to put on any greater cold-water pressure than you are in the habit of carrying. For instance, if your safety-valve is set at 110 pounds, this pressure of cold water will be sufficient to test the flues.

Now, suppose you are out in the field and want to test your flues. Of course, you have no hydrant to attach to, and you happen not to have a force pump—it would seem you were in bad shape to test your boiler with cold water. Well, you can do it by proceeding in this way: When you have expanded and beaded all the flues that were leaking, you will then close the throttle tight, take off the safety-valve (as this is generally attached at the highest point) and fill the boiler full, as it is absolutely necessary that all the space in the boiler should be filled with cold water. Then screw the safety-valve back in its place. You will then get back in the fire-box with your tools and have some one place a small sheaf of wheat or oat-straw under the fire-box and set fire to it. The expansive force of the water caused by the heat from the burning straw will produce pressure desired. You should know, however, that your safety is in perfect order. When the water begins to escape at the safety-valve, you can readily see if you have expanded your flues sufficiently to keep them from leaking.

This makes a very nice and steady pressure, and although the pressure is caused by heat, it is a cold-water pressure, as the water is not heated beyond one or two degrees. This mode of testing, however, can not be applied in very cold weather, as water has no expansive force five degrees above or five degrees below the freezing point.

This manner of testing a boiler was first employed a few years ago in the extreme Northwest by an engine expert, who was not only an expert but an artist in his line, and one who believed thoroughly in the old adage, " Where there is a will there is a way," and this was his way when he was required to test the boiler with cold water and had no pump. While this will answer the purpose, I only mention it as a means of testing when it must be done and you have no other way of doing it. A force pump is much to be preferred, as with it you can vary your pressure and hold it where you desire.

These tests, however, are only for the purpose of trying your flues and are not intended to ascertain the efficiency or strength of your boiler. When this is required, I would advise you to get an expert to do it, as the best test for this is the hammer test, and only an expert should attempt it.

PART FIFTH.

A GOOD FIREMAN.

What is a good fireman? You no doubt have heard this expression: "Where there is so much smoke there must be some fire." Well, that is true, but a good fireman doesn't make much smoke. We are now speaking of firing with coal. If I can see the smoke ten miles from a threshing engine, I can tell what kind of a fireman is running the engine; and if there is a continuous cloud of black smoke being thrown out of the smokestack, I make up my mind that the engineer is having all he can do to keep the steam up, and also conclude that there will not be much coal left by the time he gets through with the job; while, on the other hand, should I see at regular intervals a cloud of smoke going up and lasting for a few minutes, and for the next few minutes see nothing, then I conclude that the engineer of that engine knows his business, and that he is not working hard; he has plenty of steam all the time, and has coal left when he is through.

though the pressure is caused by heat, it is a cold-water pressure, as the water is not heated beyond one or two degrees. This mode of testing, however, can not be applied in very cold weather, as water has no expansive force five degrees above or five degrees below the freezing point.

This manner of testing a boiler was first employed a few years ago in the extreme Northwest by an engine expert, who was not only an expert but an artist in his line, and one who believed thoroughly in the old adage, "Where there is a will there is a way," and this was his way when he was required to test the boiler with cold water and had no pump. While this will answer the purpose, I only mention it as a means of testing when it must be done and you have no other way of doing it. A force pump is much to be preferred, as with it you can vary your pressure and hold it where you desire.

These tests, however, are only for the purpose of trying your flues and are not intended to ascertain the efficiency or strength of your boiler. When this is required, I would advise you to get an expert to do it, as the best test for this is the hammer test, and only an expert should attempt it.

The Traction Engine.

STRAW-BURNING BOILERS.

Before we leave the question of boilers we want to give you a short description of a class of boilers, which are intended to be used in places where coal or wood is expensive and where straw is practically valueless. It is not necessary to point out that you can not fire straw under a boiler intended to burn coal or even wood and expect to get good results. If straw is to be used for fuel the fire-box must be arranged for this purpose. Many manufacturers supply special fittings by which one of their coal-burning boilers can be converted into a straw burner by removing the grate-bars and substituting others intended for the burning of straw. In most cases, however, special boilers are supplied when the use of straw as a fuel is specified.

The boiler mostly used for this purpose is of the locomotive type, and is exactly like the one used for burning coal, except in regard to the fire-box.

In nearly all the fire-boxes used for this kind of fuel some kind of a shield is introduced across the back of the fire-box, in order to prevent any unburned fuel to reach the flues, and choke them, and also to insure a more perfect combustion of the fuel as well as to protect the crown-sheet from the direct flame.

On account of the very light nature of this fuel, a more

80   The Boiler.

Fig. 22.—Fire-box of the Port Huron Straw-burning Boiler.

liberal supply of air is necessary for perfect combustion, and this demand is taken care of by an additional amount of air-ducts, mostly located in the rear of the ashpit behind the grate bars. For the same reason special precautions must also be taken for the introduction of the straw into the fire-box.

The cut, Fig. 22, represents a sectional view of one of Port Huron's straw-burning boilers. The straw is introduced into the fire-box through the upper funnel, which is provided with an inward swinging door automatically closing the funnel as soon as the fuel is pushed through. The same kind of hinged door is also furnished for the ashpit. In the rear of the fire-box an opening has been provided in order to admit an additional amount of air, the quantity of which is adjusted by a sliding door. If you examine the cut carefully you will note that the front part of the grate bars nearest to the fire-door has no provisions for air-spaces. This arrangement has for object to localize the fire as much as possible in the rear of the fire-box. It also serves to catch all the chaff dropping from the straw and allow it to be burned instead of dropping directly through into the ashpit.

You will also note that a cast-iron shield or a "baffle plate" extends from the extreme end of the fire-box below the tubes nearly to the front, ending a little above the fire-door and leaving only a small space between the "baffle plate" and the front of the fire-box. The result of this arrangement is that the fuel, introduced through the above-mentioned funnel, meets the hot gases of the burnt straw, as well as an additional supply of air entering at the rear of the fire-box, and is very completely burnt, leaving very little unburnt fuel to escape into the flues. The arrows in the cut show the direction of the draught through the fire-box. The draught enters through the draught-hole below the fire funnel; some of the air passes through the grate, while a part of it goes along under and around the front end of the grates and up alongside the inner wall of the fire-box, striking the baffle plate and the newly introduced fuel. This draught of air passing under the baffle plate not only keeps it relatively cool, but also introduces fresh air where it is most useful for increasing the combustion.

In this boiler the ashes are removed through the front of the ashpit, and only air admitted from the rear or through the opening under the fuel funnel.

A small door directly above the fuel funnel allows the fireman to quickly clean the flues even when the engine is in operation.

In order to allow the flue-cleaner to reach the flues the baffle plate is cast with a box-like depression opposite this door, as can be seen from the cut.

Due to the inflammable nature of the fuel, special

precautions must be taken in order to prevent sparks or cinders to issue from the smokestack.

Most of the boilers designed for burning straw have for this reason their smoke-box made somewhat longer, as this prevents to a great extent the sparks to be blown through into the smokestack.

Various other devices are also used. They consist mostly of conical or spherical screens fitted inside or on top of the smokestack, which in the straw burners are somewhat longer than those generally used on a coal- or a wood-burning boiler.

So let us go and see what makes this difference and learn a valuable lesson. We will first go to the engine that is making such a big smoke, and we will find that the engineer has a big coal-shovel just small enough to allow it to enter the fire door. You will see the engineer throw in about two or perhaps three shovels of coal, and as a matter of course we will see a volume of black smoke issuing from the stack; the engineer stands leaning on his shovel watching the steam-gauge, and he finds that the steam does n't run up very fast, and about the time the coal gets hot enough to consume the smoke we will see him drop his shovel, pick up a poker, throw open the fire door, and commence a vigorous punching and digging at the fire. This starts the black smoke again, and about this time we will see him down on his knees with his poker, punching at the under side of the grate bars; about the time he is through with this operation the smoke is coming out less dense, and he thinks it time to throw in more coal, and he does it. Now, this is kept up all day, and you must not read this and say it is overdrawn, for it is not, and you can see it every day, and the engineer that fires in this way works hard, burns a great amount of coal, and is afraid all the time that the steam will run down on him.

Before leaving him let us take a look in his fire-box, and we will see that it is full of coal, at least up to the level of the fire door. We will also see quite a pile of ashes under the ash-pan. You can better understand the disadvantage of this way of firing after we visit the next man. I think a good way to know how to do a thing is to know, also, how not to do it.

Well, we will now go across to the man who is making but little smoke, and making that at regular intervals. We will be likely to find that he has only a little hand shovel. He picks this up, takes up a small amount of coal, opens the fire door, and spreads the coal nicely over the grates; does this quickly and shuts the door; for a minute black smoke is thrown out, but only for a minute. Why? Because he threw in only enough to replenish the fire, and not to choke it in the least, and in a minute the heat is great enough to consume all the smoke before it reaches the stack, and as smoke is unconsumed fuel, he gains that much if he can consume it. We will see this engineer standing around for the next few minutes perfectly at ease. He is not in the least afraid of his steam going down, at the end of three to five minutes, owing to the amount of work he is doing. You will see him pick up his little shovel and throw in a little coal; he does exactly as he did before, and if we stay there for an hour we will not see him pick up a poker. We will look in at his fire-box, and we will see what is called a "thin fire," but every part of the fire-box is hot. We will see but a small pile of ashes under his engine. He is not working hard.

If you happen to be thinking of buying an engine, you will say that this last fellow "has a dandy engine." "That is the kind of an engine I want," when the facts in the case may be that the first man may have a better engine, but does n't know how to fire it. Now, do n't you see how important it is that you know how to fire an engine? I am aware that some big coal wasters will say, "It is easy to talk about firing with a little hand shovel, but just get out in the field as we do and get some of the kind of fuel we have to burn, and see how you get along." Well, I am aware that you will have some bad coal. It is much better to handle bad coal in a good way than to handle good coal in a bad way. Learn to handle your fuel in the proper way and you will be a good fireman. Do n't get careless and then blame the coal for what is your own fault. Be careful about this, you might give yourself away. I have seen engineers make a big kick about the fuel and claim that it was no good, when some other fellow would take hold of the engine and have no trouble whatever. Now, this is what I call a clean give-away on the kicker.

Do n't allow any one to be a better fireman than yourself. You will see a good fireman do exactly as I have

A Good Fireman.

stated. He fires often, always keeps a level fire, never allows the coal to get up to the lower tubes, always puts in coal before the steam begins to drop, keeps the fire door open as little as possible, preventing any cold air from striking the tubes, which will not only check the steam, but is injurious to the boiler.

It is no small matter to know just how to handle your dampers; do n't allow too much of an opening here. You will keep a much more even fire by keeping the damper down, just allowing draft enough to permit free combustion; more than this is a waste of heat.

Get all out of the coal you can, and save all you get. Learn the little points that half the engineers never think of.

WOOD.

You will find wood quite different in some respects, but the good points you have learned will be useful now. Fire quick and often, but unlike coal you must keep your fire-box full. Place your wood as loosely as possible. I mean by this, place in all directions to allow the draft to pass freely all through it. Keep adding a couple of sticks as fast as there is room for them; do n't disturb the under sticks. Use short wood and fire close to the door. When firing with wood I would advise you to keep your screen down. There is much more danger of setting fire with wood than with coal.

If you are in a dangerous place, owing to the wind and the surroundings, do n't hesitate to state your fears to the man for whom you are threshing. He is not supposed to know the danger as well as you, and if, after your advice, he says go ahead, you have placed the responsibility on him; but even after you have done this it sometimes shows a good head to refuse to fire with wood, especially when you are required to fire with old rails, which is a common fuel in a timbered country. While they make a hot fire in a fire-box, they sometimes start a hot one outside of it. It is part of your business to be as careful as you can. What I mean is to take reasonable precaution, such as looking after the screen in the stack. If it burns out, get a new one. With reasonable diligence and care you will never set anything on fire, while, on the other hand, a careless engineer may do quite a lot of damage.

There is fire about an engine, and you are provided with the proper appliances to control it. See that you do it.

WHY GRATES BURN OUT.

Grates burn through carelessness. You may as well make up your mind to this at the start. You never saw grate bars burn out with a clean ash-box. They can only be burned by allowing the ashes to accumulate under them until they exclude the air, when the bars at once become red hot. The first thing they do is to warp, and if

A Good Fireman.

the ashes are not removed at once, the grate bar will burn off. Carelessness is neglecting something which is a part of your business, and as part of it is to keep your ash-box clean, it certainly is carelessness if you neglect it. Your coal may melt and run down on the bars, but if the cold air can get to the grates, the only damage this will do is to form a clinker on the top of grates, and shut off your draft. When you find that you have this kind of coal you will want to look after these clinkers.

Now, if you should have good success in keeping steam, keep improving on what you know, and if you run on 1000 pounds of coal to-day, try and do it with 900 to-morrow. That is the kind of stuff a good fireman is made of.

But do n't conclude that you can do the same amount of work each day in the week on the same amount of fuel, even if it should be of the same kind. You will find that with all your care and skill your engine will differ very materially, both as to the amount of fuel and water that it will require, though the conditions may apparently be the same.

This may be as good a time as any to say to you: Remember that a blast of cold air against the tubes is a bad thing, so be careful about your fire door; open it as little as possible; when you want to throw in fuel, do n't open the door and then go a rod away after a shovel of coal; and I will say here that I have seen this thing done by men who flattered themselves that they were about at the top in the matter of running an engine. That kind of treatment will ruin the best boiler in existence. I do n't mean that once or twice will do it, but to keep it up will do it. Get your shovel of coal, and when you are ready to throw it in, open the door quickly and close it at once. Make it one of your habits to do this, and you will never think of doing it in any other way. If it becomes necessary to stop your engine with a hot fire and a high pressure of steam, do n't throw your door open, but drop your damper and open the smoke-box door.

If, however, you expect to stop only a minute or two, drop your damper and start your injector if you have one. If you have none, get one. It is not my intention to advertise anything in this book, but I believe I can make it more valuable to you by mentioning a few things that I have tried thoroughly and found satisfactory. The Eberman Injector, made by J. Register & Sons, Baltimore, Maryland, is one of the good things for a traction engine. The Penberthy Injector, made by the Penberthy Injector Co., Detroit, Michigan, is also a good one. I shall also mention a good brand of cylinder oil, when I am ready for it, also a good reliable safety-valve, a steam-gauge, etc. These are things that you may want sometimes, and when you spend your good money you should get something reliable, and if these parties make a good article, it is due them that they be patronized. And when I say a certain article is good, I say it simply because I know it from experience, and don't mean to say that others are not good. If a friend of mine should ask me what I would advise him to use, I certainly would not advise him to get something I knew nothing of, and it is due the reader (inasmuch as he has paid for it) that I give him the benefit of my experience, that he may spend his money, when he must, where it will give him satisfaction.

THE BLOWER.

The blower is an appliance for creating artificial draft, and consists of a small pipe leading from some point above the water line into the smoke-stack, directly over the tubes, and should extend to the center of stack and terminate with a nozzle pointing directly to top and center of stack; this pipe is fitted with a globe valve. When it is required to rush your fire, you can do so by opening this globe and allowing the steam to escape into the stack. The force of the steam tends to drive the air out of the stack and the smoke-box, and create a strong draft. But, you say, "What if I have no steam?" Well, then don't blow, and be patient till you have enough to create a draft;

and it has been my experience that there is nothing gained by putting on the blower before having fifteen pounds of steam, as less pressure than this will create but little draft, and the steam will escape about as fast as it is being generated. Be patient and do n't be everlastingly punching at the fire. Get your fuel in good shape in the fire-box, and shut the door and go about your business and let the fire burn.

Must the blower be used while working the engine? No. The exhaust steam which escapes into the stack does exactly what we stated the blower does, and if it is necessary to use the blower in order to keep up steam, you can conclude that your engine is in bad shape, and yet there are times when the blower is necessary, even when your engine is in the best of condition. For instance, when you have poor fuel and are working your engine very light, the exhaust steam may not be sufficient to create enough draft for poor coal, or wet or green wood. But if you are working your engine hard the blower should never be used; if you have bad fuel and it is necessary to stop your engine, you will find it very convenient to put on the blower slightly, in order to hold your steam and keep the fire lively until you start again.

It will be a good plan for you to take a look at the nozzle on blower now and then, to see that it does not

A Good Fireman.

become limed up, and to see that it is not turned to the side so that it directs the steam to the side of stack. Should it do this, you will be using the steam and getting but little, if any, benefit. It will also be well for you to remember that you can create too much draft as well as too little. Too much draft will consume your fuel and produce but little steam.

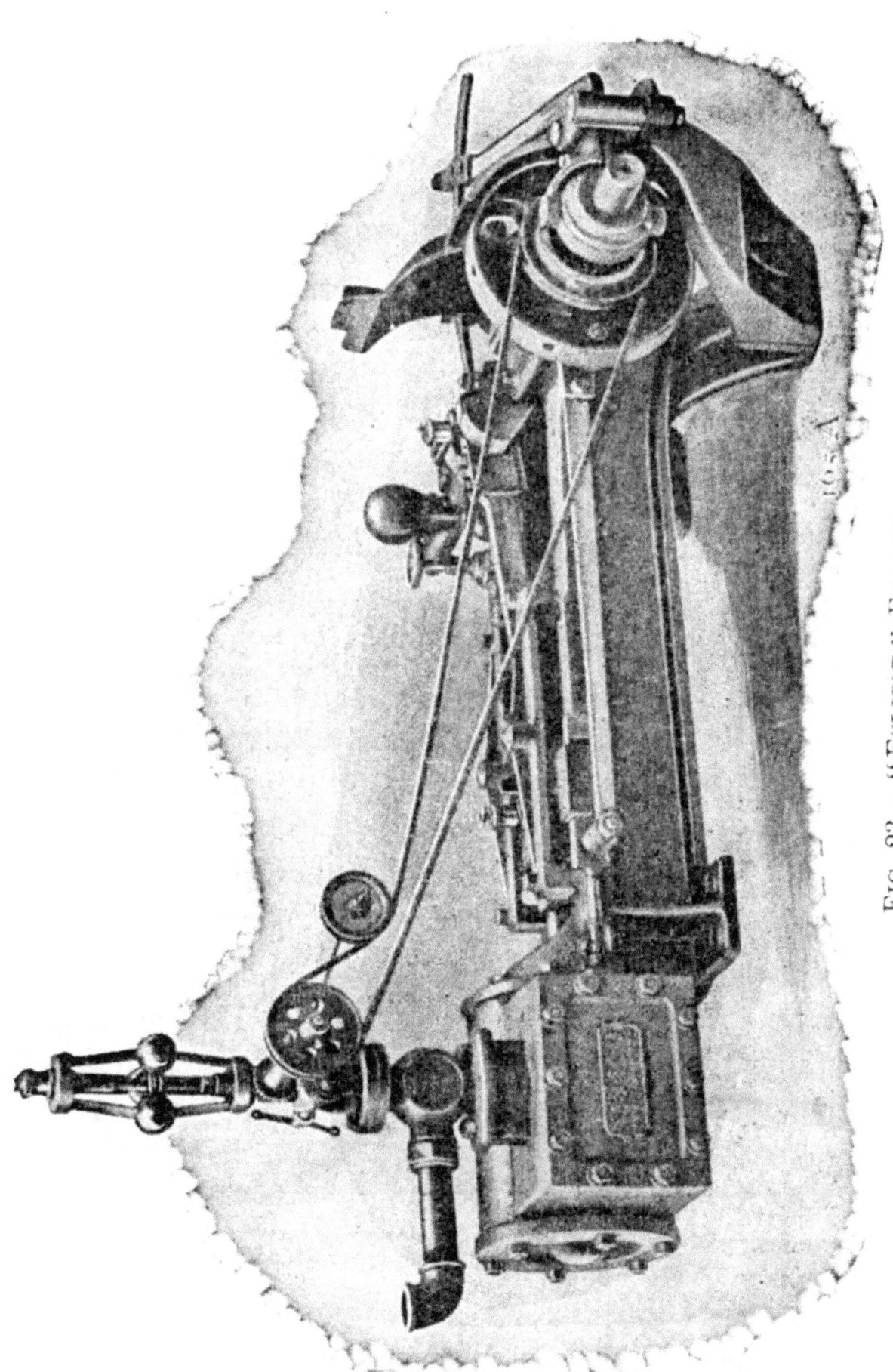

FIG. 23.—"ECLIPSE" ENGINE.

PART SIXTH.

THE ENGINE.

Any young engineer who will make use of what he has read will never get his engine into much trouble. Manufacturers of farm engines to-day make a specialty of this class of goods, and they endeavor to build them as simple and of as few parts as possible. They do this well knowing that, as a rule, they must be run by men who can not take a course in practical engineering. If each one of the many thousands of engines that are turned out every year had to have a practical engineer to run it, it would be better to be an engineer than to own the engine; and manufacturers knowing this, make their engines as simple and with as little liability to get out of order as possible. The simplest form of an engine, however, requires of the operator a certain amount of brains and a willingness to do that which he knows should be done; and if you will follow the instructions you have already received, you can run your

The Engine.

engine as successfully as any one can wish as long as your engine is in order, and, as I have just stated, it is not liable to get out of order except from constant wear, and this wear will appear in the boxes, journals, and valve. The brasses on wrist-pin and crosshead will probably require your first and most careful attention, and of these two the wrist-box will require the most; and what is true of one is true of both boxes. It is, therefore, not necessary to take up both boxes in instructing you how to handle them. We will take up the box most likely to require your attention. This is the wrist-box. You will find this box in two parts or halves. In a new engine you will find that these two halves do not meet on the wrist-pin by at least $\frac{1}{8}$ of an inch. They are brought up to the pin by means of a wedge-shaped key. (I am speaking now of the most common form of wrist-boxes. If your engine should not have this key, it will have something which serves the same purpose.) As the brasses wear you can take up this wear by forcing the key down, which brings the two halves close together. You can continue to gradually take up this wear until you have brought them together. You will then see that it is necessary to do something in order to take up any more wear, and this "something" is to take out the brasses and file about $\frac{1}{16}$ of an inch off of each brass. This will allow you another eighth of an inch to take up in wear.

Now, here is a nice little problem for you to solve and I want you to solve it to your own satisfaction, and when you do you will thoroughly understand it, and to understand it is to never allow it to get you into trouble. We started out by saying that in a new engine you would most likely find about $\frac{1}{8}$ of an inch between the brasses, and we said you would finally get these brasses or halves together and would have to take them out and file them. Now, we have taken up $\frac{1}{8}$ of an inch, and the result is we have lengthened our pitman just $\frac{1}{16}$ of an inch; or, in other words, the center of wrist-pin and the center of crosshead are just $\frac{1}{16}$ of an inch further apart than they were before any wear had taken place, and the piston head has $\frac{1}{16}$ of an inch more clearance at one end and $\frac{1}{16}$ of an inch less at the other end than it had before. Now, if we take out the boxes and file them so we have another eighth of an inch, by the time we have taken up this wear we will then have this distance doubled, and we will soon have the piston head striking the end of the cylinder, and, besides, the engine will not run as smooth as it did. Half of the wear comes off of each half, and the half next to the key is brought up to the wrist-pin because of the tapering key, while the outside half remains in one place. You must therefore place back of this half a thin piece of sheet copper, or a piece of tin will do. Now, suppose our boxes had $\frac{1}{8}$ of an inch for wear; when we have taken up this

The Engine.

much we must put in $\frac{1}{16}$ of an inch backing (as it is called), for we have reduced the outside half by just that amount. We have also reduced the front half the same, but, as we have said, the tapering key brings this half up to its place.

Now, we think we have made this clear enough and we will leave this and go back to the key again. You must remember that we stated that the key was tapering or a wedged-shape, and, as a wedge, is equally as powerful as a screw; and you must bear in mind that a slight tap will bring these two boxes up tight against the wrist-pin. Young engineers experience more trouble with this box than with any other part of the engine, and all because they do not know how to manage it. You should be very careful not to get your box too tight, and don't imagine that every time there is a little knock about your engine that you can stop it by driving the key down a little more. This is a great mistake that many, and even old, engineers make. I at one time saw a wrist-pin and boxes ruined by the engineer trying to stop a knock that came from a loose fly-wheel. It is a fact, and one that has never been satisfactorily explained, that a knock coming from almost any part of an engine will appear to be in the wrist. So bear this in mind and don't allow yourself to be deceived in this way, and never try to stop a knock until you have first located the trouble beyond a doubt.

When it becomes necessary to key up your brasses, you will find it a good safe way to loosen up the set screw which holds the key, then drive it down till you are satisfied you have it tight. Then drive it back again and then with your fist drive the key down as far as you can. You may consider this a peculiar kind of a hammer, but your boxes will rarely ever heat after being keyed in this manner.

KNOCK IN ENGINES.

What makes an engine knock or pound? A loose pillow block box is a good "knocker." The pillow block is a box next crank or disc wheel. This box is usually fitted with set bolts and jam nuts. You must also be careful not to set this up too tight, remembering always that a box when too tight begins to heat, and this expands the journal, causing greater friction. A slight turn of a set bolt one way or the other may be sufficient to cool a box that may be running hot, or to heat one that may be running cool. A hot box from neglect of oiling can be cooled by supplying oil, provided it has not already commenced to cut. If it shows any sign of cutting, the only safe way is to remove the box and clean it thoroughly.

Loose eccentric yokes will make a knock in an engine, and it may appear to be in the wrist. You will find packing between the two halves of the yoke. Take out a thin sheet of this packing, but do n't take out too much, as you

The Engine.

are liable then to get them too tight, and they may stick and cause your eccentrics to slip. We will have more to say about the slipping of the eccentrics.

The piston rod loose in crosshead will make a knock, which also appears in the wrist, but it is not there. Tighten the piston and you will stop it.

The crosshead loose in the guides will make it knock. If the crosshead is not provided for taking up this wear, you can take off the guides and file them enough to allow them to come up to the crosshead, but it is much better to have them planed off, which insures the guides coming up square against the crosshead and thus prevent any heating or cutting.

A loose fly-wheel will most likely puzzle you more than anything else to find the knock. So remember this: The wheel may apparently be tight, but should the key be the least bit narrow for the groove in shaft, it will make your engine bump very similar to that caused by too much or too little "lead."

LEAD.

What is lead? Lead is a space or opening of port on steam end of cylinder when engine is on dead center (dead center is the two points of disc or crank wheel at which the crank pin is in direct line with piston, and at which no amount of steam will start the engine). Different makes of engines differ to such an extent that it is impossible to give any rule or any definite amount of lead for an engine. For instance, an engine with a port six inches long and $\frac{1}{2}$ of an inch wide would require much less lead than one with a port four inches long and one inch wide. Suppose I should say $\frac{1}{16}$ of an inch was the proper lead. In one engine you would have an opening $\frac{1}{16}$ of an inch wide and six inches long, and in the other you would have $\frac{1}{16}$ of an inch wide and four inches long; so you can readily see that it is impossible to give the amount of lead for an engine. Lead allows live steam to enter the cylinder just ahead of the piston at the point of finishing the stroke, and forms a "cushion," and enables the engine to pass the center without a jar. Too much lead is a source of weakness to an engine, as it allows the steam to enter the cylinder too soon and forms a back pressure, and tends to prevent the engine from passing the center. It will, therefore, make your engine bump, and make it very difficult to hold the packing in stuffing-box.

Insufficient lead will not allow enough steam to enter the cylinder ahead of piston to afford cushion enough to stop the inertia, and the result will be that your engine will pound on the wrist-pin. You most likely have concluded by this time that "lead" is no small factor in the smooth running of an engine, and you, as a matter of course, will want to know how you are to obtain the proper

The Engine.

lead. Well, don't worry yourself. Your engine is not going to have too much lead to-day and not enough to-morrow. If your engine was properly set up in the first place, the lead will be all right, and continue to afford the proper lead as long as the valve has not been disturbed from its original position; and this brings us to the most important duty of an engineer as far as the engine is concerned, viz., setting the valve.

SETTING A VALVE.

The proper and accurate setting of a valve on a steam engine is one of the most important duties that you will have to perform, as it requires a nicety of calculation and a mechanical accuracy. And when we remember also that this is another one of the things for which no uniform rule can be adopted, owing to the many circumstances which go to make an engine so different under different conditions, we find it very difficult to give you the light on this part of your duty which we would wish to. We, however, hope to make it so clear to you that by the aid of the engine before you you can readily understand the conditions and principles which control the valve in the particular engine which you may have under your management.

The power and economy of an engine depend largely on the accurate operation of its valve. It is, therefore, necessary that you know how to reset it, should it become necessary to do so.

An authority says, "Bring your engine to a dead center, and then adjust your valve to the proper lead." This is all right as far as it goes, but how are you to find the dead center? I know that it is a common custom in the field to bring the engine to a center by the use of the eye. You may have a good eye, but it is not good enough to depend on for the accurate setting of a valve.

HOW TO FIND THE DEAD CENTER.

First, provide yourself with a "tram." This you can do by taking a $\frac{1}{4}$ inch iron rod, about eighteen inches long, and bend about two inches of one end to a sharp angle. Then sharpen both ends to a nice sharp point. Now fasten securely a block of hard wood somewhere near the face of the fly-wheel, so that when the straight end of your tram is placed at a definite point in the block, the other, or hook end, will reach the crown of the fly-wheel.

Be certain that the block can not move from its place, and be careful to place the tram at exactly the same point on the block at each time you bring the tram into use. You are now ready to proceed to find the dead center, and in doing this remember to turn the fly-wheel always in the same direction. Now turn your engine over until

The Engine.

it nears one of the centers, but not quite to it. You will then, by the aid of a straight-edge, make a clear and distinct mark across the guides and crosshead. Now go around to the fly-wheel and place the straight end of the tram at some point on the block, and with the hook end make a mark across the crown or center of face of fly-wheel; now turn your engine past the center and on to the point at which the line on crosshead is exactly in line with the lines on guides. Now place your tram in the same place as before, and make another mark across the crown of the fly-wheel. By the use of dividers find the exact center between the two marks made on fly-wheel; mark this point with a center-punch. Now bring the fly-wheel to the point at which, when the tram is placed at its proper place on block, the hook end, or point, will touch this punch-mark, and you will have one of the exact dead centers.

Now turn the engine over until it nears the other center, and proceed exactly as before, remembering always to place the straight end of tram exactly in same place in block, and you will find both dead centers as accurately as if you had all the fine tools of an engine-builder.

You are now ready to proceed with the setting of your valve, and as you have both dead centers to work from, you ought to be able to do it, as you do not have to depend on your eye to find them, and by the use of the tram you turn your engine to exactly the same point every time you wish to get a center.

Now remove the cap on steam-chest, bring your engine to a dead center, and give your valve the necessary amount of lead on the steam end. Now, we have already stated that we could not give you the proper amount of lead for an engine. It is presumed that the maker of your engine knew the amount best adapted to this engine, and you can ascertain his idea of this by first allowing, we will say, about $\frac{1}{16}$ of an inch. Now bring your engine to the other center, and if the lead at the other part is less than $\frac{1}{16}$, then you must conclude that he intended to allow less than $\frac{1}{16}$; but should it show more than this, then it is evident that he intended more than $\frac{1}{16}$ lead; but in either case you must adjust your valve so as to divide the space, in order to secure the same lead when on either center. In the absence of any better tool to ascertain if the lead is the same, make a tapering wooden wedge of soft wood, turn the engine to a center and force the wedge in the opening made by the valve hard enough to mark the wood; then turn to the next center, and if the wedge enters the same distance, you are correct; if not, adjust till it does, and when you have it set at the proper place, you had best mark it by taking a sharp cold chisel and place it so that it will cut into the hub of eccentric and

The Engine.

in the shaft; then hit it a smart blow with a hammer. This should be done after you have set the set screws in eccentric down solid on the shaft. Then, at any time should your eccentric slip, you have only to bring it back to the chisel mark and fasten it and you are ready to go ahead again.

This is for a plain or single eccentric engine. A double or reversible engine, however, is somewhat more difficult to handle in setting the valve. Not that the valve itself is any different from a plain engine, but from the fact that the link may confuse you; and while the link may be in position to run the engine one way, you may be endeavoring to set the valve to run it the other way.

The proper way to proceed with this kind of an engine is to bring the reverse lever to a position to run the engine forward, then proceed to set your valve the same as on a plain engine. When you have it at the proper place, tighten just enough to keep from slipping, then bring your reverse lever to the reverse position and bring your engine to the center. If it shows the same lead for the reverse motion, you are then ready to tighten your eccentrics securely, and they should be marked as before.

You may imagine that you will have this to do often. Well, don't be scared about it. You may run an engine a long time, and never have to set a valve. I have heard these windy engineers (you have seen them) say that they had to go and set Mr. A's or Mr. B's valve, when the facts were, if they did anything, it was simply to bring the eccentrics back to their original position. They happened to know that almost all engines are plainly marked at the factory, and all there was to do was to bring the eccentrics back to these marks and fasten them, and the valve was set. The slipping of the eccentrics is about the only cause for a valve working badly. You should, therefore, keep all grease and dirt away from these marks; keep the set-screws well tightened, and notice them frequently to see that they do not slip. Should they slip a $\frac{1}{16}$ part of an inch, a well-educated ear can detect it in the exhaust. Should they slip a part of a turn, as they will sometimes, the engine may stop instantly, or it may cut a few peculiar circles for a minute or two, but don't get excited; look to the eccentrics at once for the trouble.

Your engine may, however, act very queerly some time, and you may find the eccentrics in their proper place. Then you must go into the steam-chest for the trouble. The valves in different engines are fastened on the valve rod in different ways. Some are held in place by jam nuts; a nut may have worked loose, causing lost motion on the valve. This will make your engine work badly. Other engines hold their valve by a clamp and pin. This pin may work out, and when it does your engine will stop, and stop very quickly, too.

The Engine.

If you thoroughly understand the working of the steam, you can readily detect any defect in your cylinder or steam-chest by the use of your cylinder cocks. Suppose we try them. Turn your engine on the forward center; now open the cocks and give the engine the steam pressure. If the steam blows out at the forward cock, we know that we have sufficient lead. Now turn back to the back center, and give it steam again; if it blows out the same at this cock, we can conclude that our valve is in its proper position. Now reverse the engine and do the same thing; if the cocks act the same, we know we are right. But suppose the steam blows out of one cock all right, and when we bring the engine to the other center no steam escapes from this cock, then we know that something is wrong with the valve; and if the eccentrics are in their proper position, the trouble must be in the steam-chest, and if we open it up we shall find the valve has become loosened on the rod. Again, suppose we put the engine on a center, and, on giving it steam, we find the steam blowing out at both cocks.

Now, what is the trouble?—for no engine in perfect shape will allow the steam to blow out of both cocks at the same time. It is one of two things, and it is difficult to tell. Either the cylinder rings leak and allow the steam to blow through, or else the valve is cut on the seat, and allows the steam to blow over. Either of these two causes is bad, as it not only weakens your engine, but is a great waste of fuel and water. The way to determine which of the two causes this is to take off the cylinder head, turn engine on forward center, and open throttle slightly. If the steam is seen to blow out of the port at open end of cylinder, then the trouble is in the valve; but if not, you will see it blowing through from forward end of cylinder, and the trouble is in the cylinder rings.

What is the remedy? Well, if the "rings" are the trouble, a new set will most likely remedy it should they be of the automatic or self-setting pattern; but should they be of the spring or adjusting pattern, you can take out the head and set the rings out to stop this blowing. As nearly all engines now are using the self-setting rings, you will most likely require a new set.

If the trouble is in the valve or steam-chest, you had better take it off and have the valve seat planed down and the valve seated to it. This is the safest and best way. Never attempt to dress a valve down; you are almost certain to make a bad job of it.

LUBRICATING OIL.

What is oil?

Oil is a coating for a journal, or, in other words, is a lining between bearings.

Did you ever stop long enough to ask yourself the

question? I doubt it. A great many people buy something to use on their engine because it is called oil. Now, if the object in using oil is to keep a lining between the bearings, is it not reasonable to use something that will adhere to that which it is to line or cover?

Gasoline will cover a journal for a minute or two; an oil a grade better would last a few minutes longer. Still another grade would do somewhat better. Now, if you are running your own engine, buy the best oil you can buy. You will find it very poor economy to buy cheap oil; and if you are not posted, you may pay price enough but get a very poor article.

If you are running an engine for some one else, make it part of your contract that you are furnished with a good oil. You can not keep an engine in good shape with a cheap oil. You say "you are going to keep your engine clean and bright." Not if you must use a poor oil.

Well, how are you to know when you are getting good oil? The best way is to ascertain a good brand and then use that and nothing else. We are not selling or advertising oil, but if you use the "Eldorado Castor Oil" on your engine you will have the best, or as good as you can buy. How will it work in the cylinder? Well, it will not work there; it is not intended for cylinder, as it will not stand the heat. If you are carrying ninety pounds of steam, you will have about 320 degrees of heat in cylinder; with 120 pounds, you have 341 degrees.

Now, if you want a lining between your valves and valve seat, and between your piston head and surface of cylinder, you want something that will not only stand this heat, but stand considerable more, so that it will have some staying qualities—"Capital Cylinder Oil" will do it. If your link has been knocking, just try this, and if it does n't stop, it will be because you have some connections that want attention, and want it badly.

I think "Helmet Oil," as a solid lubricant, is a great oil for what it is intended. I can't tell you who makes it, but your dealer can get it for you.

The only objection to the grease is that it requires a cup adapted to its use. Charles I. Beasley & Co., of Chicago, manufacture a good cup. While it is not automatic, I think when you once understand the nature of this kind of lubricant you would not pay the difference in price between this cup and the automatic cup.

In attaching these grease-cups on boxes not previously arranged for them, it would be well for you to know how to do it properly. You will remove the journal, take a gauge, and cut a clean groove across the box, starting in at one corner, about $\frac{1}{8}$ of an inch from the point of box, and cut diagonally across, coming out at the opposite corner on the other end of box. Then start at the opposite corner and run through as before, crossing the first groove in the center of box. Groove both halves of box alike, being careful not to cut out at either end, as this will allow the grease to escape from box and cause unnecessary waste. The chimney or packing in box should be cut so as to touch the journal at both ends of box, but not in the center or between these two points. So, when the cup is brought down tight, this will form another reservoir for the grease. If the box is not tapped directly in the center for cup, it will be necessary to cut other grooves from where it is tapped into the grooves already made. A box prepared in this way will require but little attention if you use good grease.

A HOT BOX.

You will sometimes get a hot box. What is the best remedy? Well, I might name you a dozen, and if I did you would most likely never have one on hand when it was wanted. So I will only give you one, and that is white lead and oil, and I want you to provide yourself with a can of this useful article. And should a journal or box get hot on your hands and refuse to cool with the usual methods, remove the cup, and, after mixing a portion of the lead with oil, put a heavy coat of it on the journal, put back the cup, and your journal will cool off very quickly. Be careful to keep all grit or dust out of your can of lead. Look after this part of it yourself. It is your business.

PART SEVENTH.

HANDLING A TRACTION ENGINE.

Before taking up the handling of a traction engine we want to tell you of a number of things you are likely to do which you ought not to do.

Don't open the throttle too quickly, or you may throw the drive belt off, and are also more apt to raise the water and start priming.

Don't attempt to start the engine with the cylinder cocks closed, but make it a habit to open them when you stop; this will always insure your cylinder being free from water on starting.

Don't talk too much while on duty.

Don't pull the ashes out of ash-pan unless you have a bucket of water handy.

Don't start the pump when you know you have low water.

Don't let it get low.

Don't let your engine get dirty.

Don't say you can't keep it clean.

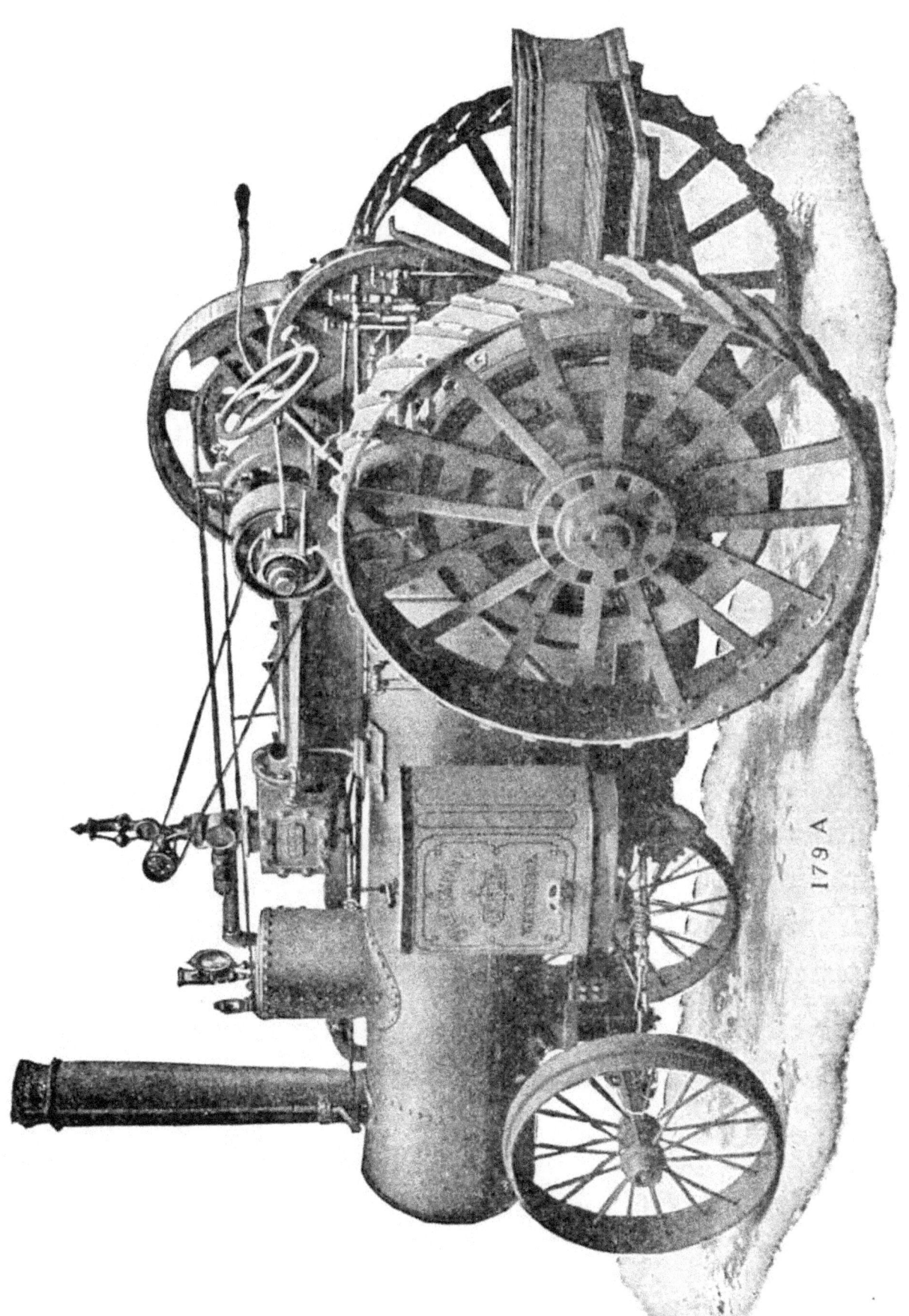

Fig. 24.—The "Eclipse" Traction Engine.

116

Handling a Traction Engine.

Do n't leave your engine at night until you have covered it up.

Do n't let the exhaust nozzle lime up, and do n't allow lime to collect where the water enters the boiler, or you may split a heater pipe or knock the top off of a check-valve.

Do n't leave your engine in cold weather without first draining all pipes.

Do n't disconnect your engine with a leaky throttle.

Do n't allow the steam to vary more than ten or fifteen pounds while at work.

Do n't allow any one to fool with your engine.

Do n't try any foolish experiments on your engine.

Do n't run an old boiler without first having it thoroughly tested.

Do n't stop when descending a steep grade.

Do n't pull through a stockyard without first closing the damper tight.

Do n't pull on to a strange bridge without first examining it.

Do n't run any risk on a bad bridge.

A TRACTION ENGINE.

You may know all about an engine. You may be able to build one, and yet run a traction into the ditch the first jump.

It is a fact that some men never can become good operators of a traction engine, and I can't give you the reason why, any more than you can tell why one man can handle a pair of horses better than another man who has had the same advantages. And yet, if you do ditch your engine a few times, don't conclude that you can never handle a traction.

If you are going to run a traction engine I would advise you to use your best efforts to become an expert at it. For the expert will hook up to his load and get out of the neighborhood while the awkward fellow is getting his engine around ready to hook up.

The expert will line up to the separator for the first time, while the other fellow will back and turn around for half an hour, and then not have a good job.

Now, don't make the fatal mistake of thinking that the fellow is an expert who jumps up on his engine and jerks the throttle open and yanks it around backward and forward, reversing with a snap, and makes it stand up on its hind wheels.

If you want to be an expert, you must begin with the throttle; therein lies the secret of the real expert. He feels the power of his engine through the throttle. He opens it just enough to do what he wants it to do. He therefore has complete control of his engine. The fellow who backs his engine up to the separator with an open

Handling a Traction Engine.

throttle and must reverse it to keep from running into and breaking something, is running his engine on his muscle and is entitled to small pay.

The expert brings his engine back under full control, and stops it exactly where he wants it. He handles his engine with his head and should be paid accordingly. He never makes a false move, loses no time, breaks nothing, makes no unnecessary noise, doesn't get the water all stirred up in the boiler, hooks up, and moves out in the same quiet manner, and the onlookers think he could pull two such loads, and say he has a great engine; while the engineer of muscle would back up and jerk his engine around a half dozen times before he could make the coupling, then with a jerk and a snort he yanks the separator out of the holes, and the onlookers think he has about all he can pull.

Now, these are facts, and they can not be put too strongly, and if you are going to depend on your muscle to run your engine, don't ask any more money than you would get at any other day labor.

You are not expected to become an expert all at once. Three things are essential to be able to handle a traction engine as it should be handled.

First, a thorough knowledge of the throttle. I don't mean that you should simply know how to pull it open and shut it. Any boy can do that. But I mean that you should be a good judge of the amount of power it will require to do what you may wish to do, and then give it the amount of throttle that it will require and no more. To illustrate this I will give you an instance.

An expert was called a long distance to see an engine that the operator said would not pull its load over the hills he had to travel.

The first pull he had to make was up the worst hill he had. When he approached the grade he threw off the governor belt, opened the throttle as wide as he could get it, and made a run for the hill. The result was that he lifted the water and choked the engine down before he was half way up. He stepped off with the remark, "That is the way the thing does." The expert then locked the hind wheels of the separator with a timber, and, without raising the pressure a pound, pulled it over the hill. He gave it just throttle enough to pull the load, and made no effort to hurry it, and still had power to spare.

A locomotive engineer makes a run for a hill in order that the momentum of his train may help carry him over. It is not so with a traction and its load; the momentum that you get does n't push very hard.

The engineer who does n't know how to throttle his engine never knows what it will do, and therefore has but little confidence in it; while the engineer who has a

Handling a Traction Engine.

thorough knowledge of the throttle and uses it, always has power to spare and has perfect confidence in his engine. He knows exactly what he can do and what he can not do.

The second thing for you to know is to get on to the tricks of the steer-wheel. This will come to you naturally, and it is not necessary for me to spend much time on it. All new beginners make the mistake of turning the wheel too often. Remember this—that every extra turn to the right requires two turns to the left, and every extra turn to the left requires two more to the right; especially is this the case if your engine is fast on the road.

The third thing for you to learn is to keep your eyes on the front wheels of your engine, and not be looking back to see if your load is coming.

In making a difficult turn, you will find it very much to your advantage to go slow, as it gives you much better control of your front wheels, and it is not a bad plan for a beginner to continue to go slow until he has perfect confidence in his ability to handle the steer-wheel, as it may keep him out of some bad scrapes.

How about getting into a hole? Well, you are not interested half as much in knowing how to get into a hole as you are in knowing how to get out. An engineer never shows the stuff he is made of to such good advantage as when he gets into a hole; and he is sure to get

The Traction Engine.

there, for one of the traits of a traction engine is its natural ability to find a soft place in the ground.

Head work will get you out of a bad place quicker than all the steam you can get in your boiler. Never allow the drivers to turn without doing some good. If you are in a hole, and you are able to turn your wheels, you are not stuck; but do n't allow your wheels to slip, it only lets you in deeper. If your wheels can't get a footing, you want to give them something to hold to. Most smart engineers will tell you that the best thing is a heavy chain. That is true. So are gold dollars the best things to buy bread with, but you have not always got the gold dollars; neither have you always got the chain. Old hay or straw is a good thing; old rails or timber of any kind. The engineer with a head spends more time trying to get his wheels to hold than he does trying to pull out, while the one without a head spends more time trying to pull out than he does trying to secure a footing; and the result is that the first fellow generally gets out the first attempt, while the other fellow is lucky if he gets out the first half-day.

If you have one wheel perfectly secure, do n't spoil it by starting your engine until you have the other just as secure.

If you get into a place where your engine is unable to turn its wheels, then you are stuck, and the only thing for you to do is to lighten your load or dig out. But under all circumstances your engine should be given the benefit of your brain.

All traction engines to be practical must, of necessity, be reversible. To accomplish this, the link with the double eccentric is largely used, although various other devices are used with success. As they all accomplish the same purpose, it is not necessary for us to discuss the merits or demerits of either.

The main object is to enable the operator to run his engine either backward or forward at will; but the link is also a great cause of economy, as it enables the engineer to use the steam more or less expansively, as he may use more or less power, and especially is this true while the engine is on the road, as the power required may vary, in going a short distance, anywhere from nothing in going down hill, to the full power of your engine in going up.

By using steam expansively we mean the cutting off of the steam from the cylinder when the piston has traveled a certain part of its stroke. The earlier in the stroke this is accomplished, the more benefit you get of the expansive force of the steam.

The reverse on traction engines is usually arranged to cut off at $\frac{1}{4}$, $\frac{1}{2}$, or $\frac{3}{4}$. To illustrate what is meant by "cutting off" at $\frac{1}{4}$, $\frac{1}{2}$, or $\frac{3}{4}$, we will suppose the engine has a twelve-inch stroke. The piston begins its stroke at the end of cylinder, and is driven by live steam through an open port three inches, or one-quarter of the stroke, when the port is closed by the valve shutting the steam from the boiler, and the piston is driven the remaining nine inches of its stroke by the expansive force of the steam. By cutting off at $\frac{1}{2}$ we mean that the piston is driven half its stroke, or six inches, by live steam, and by the expansion of the steam the remaining six inches; by $\frac{3}{4}$ we mean that live steam is used nine inches before cutting off, and expansively the remaining three inches of stroke.

Here is something for you to remember: "The earlier in the stroke you cut off, the greater the economy, but less the power; the later you cut off, the less the economy and greater the power."

Suppose we go into this a little farther. If you are carrying 100 pounds pressure and cut off at $\frac{1}{4}$, you can readily see the economy of fuel and water, for the steam is only allowed to enter the cylinder during $\frac{1}{4}$ of its stroke; but by reason of this, you only get an average pressure on the piston head of fifty-nine pounds throughout the stroke. But if this is sufficient to do the work, why not take advantage of it and thereby save your fuel and water? Now, with the same pressure as before, and cutting off at $\frac{1}{2}$, you have an average pressure on piston head of eighty-four pounds, a loss of fifty per cent.

Handling a Traction Engine.

in economy and a gain of forty-two per cent. in power. Cutting off at $\frac{3}{4}$ gives you an average pressure of ninety-six pounds throughout the stroke. A loss on cutting off at $\frac{1}{4}$ of seventy-five per cent. in economy, and a gain of nearly sixty-three per cent. in power, show that the most available point at which to work steam expansively is at $\frac{1}{4}$, as the percentage of increase of power does not equal the percentage of loss in economy. The nearer you bring the reverse lever to center of quadrant, the earlier will the valve cut the steam and less will be the average pressure, while the farther away from the center, the later in the stroke will the valve cut the steam, and the greater the average pressure, and, consequently, the greater the power. We have seen engineers drop the reverse back in the last notch in order to make a hard pull, and were unable to tell why they did so.

Now, so far as doing the work is concerned, it is not absolutely necessary that you know this; but if you do know it, you are more likely to profit by it and thereby get the best results out of your engine. And as this is our object, we want you to know it, and be benefited by the knowledge. Suppose you are on the road with your engine and load, and you have a stretch of nice road. You are carrying a good head of steam and running with lever back in the corner or lower notch. Now, your engine will travel along its regular speed, and say you run a mile this way and fire twice in making it. You now ought to be able to turn around and go back on the same road with one fire by simply hooking the lever up as short as will allow to do the work. Your engine will make the same time with half the fuel and water, simply because you utilize the expansive force of the steam instead of using the live steam from boiler. A great many good engines are condemned and said to use too much fuel, and all because the engineer takes no pains to utilize the steam to the best advantage.

I have already advised you to carry a "high pressure"—by a high pressure I mean anywhere from 100 to 120 pounds. I have done this expecting you to use the steam expansively whenever possible, and the expansive force of steam increases very rapidly after you have reached seventy pounds. Steam at eighty pounds used expansively will do nine times the work of steam at twenty-five pounds. Note the difference. Pressure, three and one-fifth times greater. Work performed, nine times greater. I give you these facts trusting that you will take advantage of them; and if your engine at 100 or 110 pounds will do your work, cutting off at $\frac{1}{4}$, don't allow it to cut off at $\frac{1}{2}$. If cutting off at $\frac{1}{2}$ will do the work, don't allow it to cut off at $\frac{3}{4}$, and the result will be that you will do the work with the least possible amount of fuel, and no one will have any reason to find fault with you or your engine.

Handling a Traction Engine.

Now we have given you the three points which are absolutely necessary to the successful handling of a traction engine. We went through it with you when running as a stationary; then we gave you the pointers to be observed when running as a traction or road engine. We have also given you hints on economy, and if you do not already know too much to follow our advice, you can get into the field with an engine and have no fears as to the results.

How about bad bridges?

Well, a bad bridge is a bad thing, and you can not be too careful. When you have questionable bridges to cross over, you should provide yourself with good hard-wood planks. If you can, have them sawed to order, have them three inches in the center and tapering to two inches at the ends. You should have two of these about sixteen feet long, and two 2 x 12 planks about eight feet long; the short ones for culverts, and for helping with the longer ones in crossing longer bridges.

An engine should never be allowed to drop from a set of planks down on to the floor of bridge. This is why I advocate four planks. Don't hesitate to use the plank. You had better plank a dozen bridges that don't need it than to attempt to cross one that does need it. You will also find it very convenient to carry at least fifty feet of good heavy rope. Don't attempt to pull across a doubtful bridge with the separator or tank hooked directly to the engine. It is dangerous. Here is where you want the rope. An engine should be run across a bad bridge very slowly and carefully, and not allowed to jerk. In extreme cases it is better to run across by hand; don't do this but once; get after the road supervisors.

SAND.

An engineer wants a sufficient amount of "sand," but he does n't want it in the road. However, you will find it there, and it is the meanest road you will have to travel. A bad sand road requires considerable sleight of hand on the part of the engineer if he wishes to pull much of a load through it. You will find it to your advantage to keep your engine as straight as possible, as you are not so liable to start one wheel to slipping any sooner than the other. Never attempt to "wiggle" through a sand bar, and do n't try to hurry through; be satisfied with going slow, just as long as you are going. An engine will stand a certain speed through sand, and the moment you attempt to increase that speed you break its footing, and then you are gone. In a case of this kind, a few bundles of hay is about the best thing you can use under your drivers in order to get started again. But do n't lose your temper; it won't help the sand any.

PART EIGHTH.

DIFFERENT TRACTION ENGINES.

In the following pages I will give you a short description of the more prominent manufacturers' latest designs in traction engines.

THE "ECLIPSE" TRACTION ENGINE.

In this engine, manufactured by the Frick Co., a separate steel frame has been provided for the support of the driving wheels, the front wheels, the engine, etc., and on this frame rests also the boiler, supported on sliding joints. By this arrangement the boiler is relieved not only of all outside strains due to the mounting of the machinery directly on the boiler shell, but also of internal strains due to the expansion or contraction of its various parts.

The boiler, of the regular locomotive type, is provided with a large fire-box and an ample supply of tubes. The crown-sheet is so arranged that it is well under the surface of the water whether the engine is going up a steep grade or going down.

The Traction Engine.

The smokestack is located in the front, and is provided with a cone-shaped screen for spark arrester. A cross-head pump and a Penberthy injector are used. The water-tanks, one on each side of the boiler, are supported directly on the steel frame and located in front of the driving-wheel.

The engine, of the center crank type, is placed on top of the boiler. It is independently supported on a steel frame spanning the boiler and fire-box and riveted to the side of the main frame. The cylinder end rests loosely on the shell of the boiler. Cylinder, bed-plate, cross-head, and slides are made of cast-iron; and cross-head gibs, crank-shaft bearings, check-valves and connecting-rod boxes are gun metal. The gears are all cast-iron. Steel is used for the main axle, the countershaft, the crank shaft, the piston rod, the pump plunger and the valve stems.

The power is transmitted from the engine to the main axle by means of a direct train of gears in which is included a compensating gear. The compensating gear is so arranged that by pulling on a lever the two driving-wheels are locked together. It is hardly necessary to point out the value of this device when pulling out of a bad place.

The connection between the main gears and the driving-wheels is accomplished by means of four spiral steel

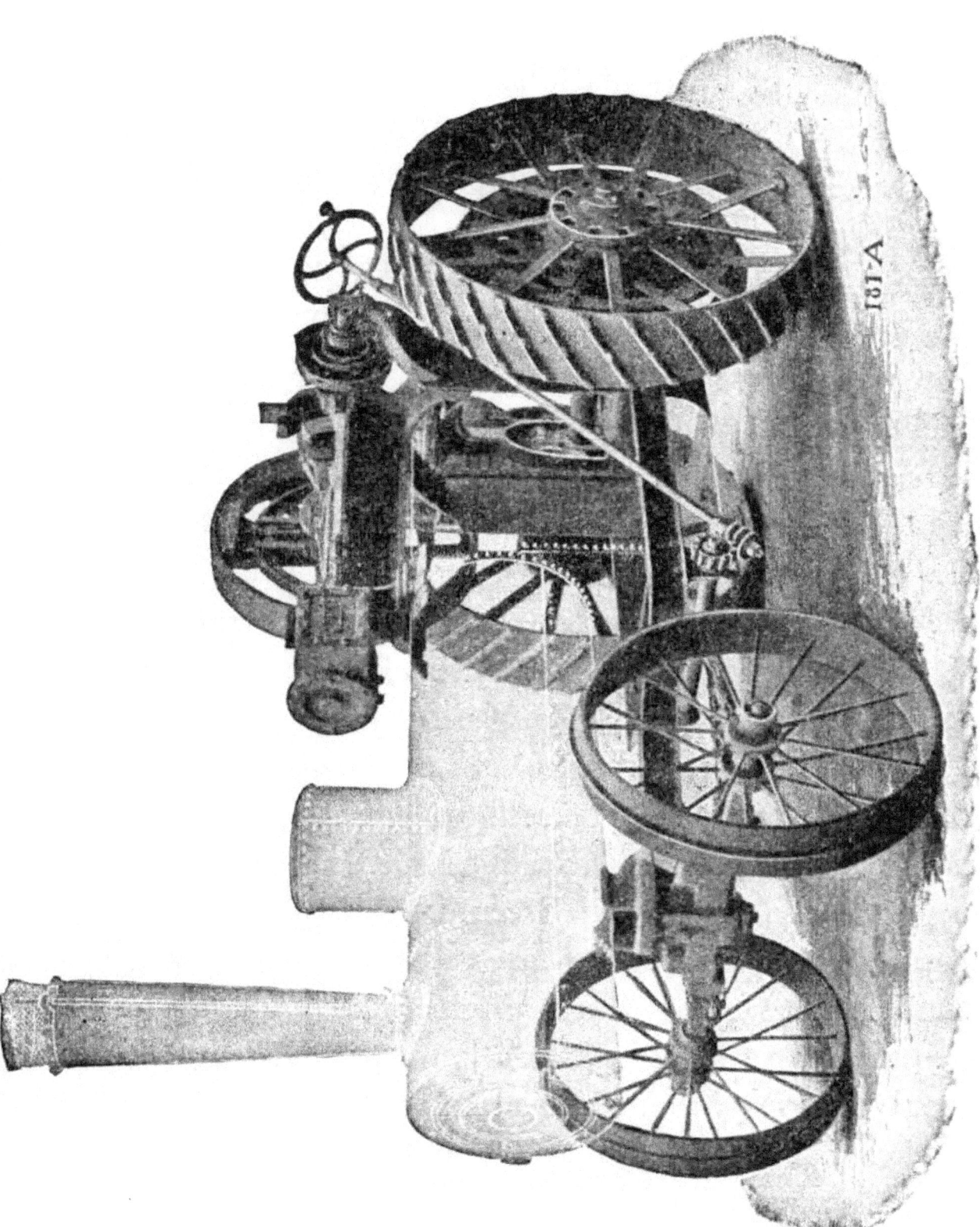

FIG. 25.—THE "ECLIPSE" TRACTION ENGINE.

springs in order to prevent as much as possible any sudden blows on gears and shafts.

In place of the usual locomotive links a special reversing gear is used. It is of a simple construction and contains very few parts.

Most of the wearing parts are encased in dust-free covers, and provided with self-oiling devices. The cylinder is oiled by a sight-feed lubricator.

The steering chains operated by the steering wheel and a worm gear are attached to the front axle by means of spiral steel springs. This way of attaching the chains allows the front axle to move a little when a small obstacle in the road is encountered, but compels it to resume its original position as soon as the obstacle has been passed. It also prevents a great deal of vibration.

The brake is applied directly on the main axle and is of the steel band type. The fly-wheel is provided with a friction clutch and the location is such that the belt can be run to the front as well as to the rear of the machine. The engineer's platform is mounted on springs. Steering wheel brake, reversing gear, throttle valve, whistle, pump, injector, blower, water-gauge, friction clutch, dampers, etc., can all be reached from this place.

The traction wheels' spokes are made of rolled steel and riveted to the rim. Holes are provided in the rim for insertion of special spurs in order to prevent slipping on frozen roads, or can be used for the attachment of special large mud cleats in case such should be wanted.

THE AULTMAN CO.'S TRACTION ENGINE.

This engine has, like the previous one, the boiler and all the machinery supported on an independent steel frame built up of structural steel consisting mostly of I beams riveted together. As shown in the cuts, see Figs. 26 and 27, the engines have been placed directly on this frame in front of the fire-box and underneath the boiler. This construction not only relieves the boiler of all external strains, but also brings the center of gravity of the whole machine closer to the road; that is, makes it less top-heavy, which is a very good feature for any traction engine.

The front end of the frame is connected with the fifth wheel or "saddle" by means of arched girders or "goose necks," allowing the front wheels to turn completely around if wanted and also giving to the whole frame a certain amount of elasticity which is very desirable.

In order to reduce the work of steering the front axle "saddle" has been furnished with ball bearings.

The placing of the front wheels ahead of the boiler has necessarily produced a somewhat longer wheel base than if the boiler itself had been used for the support of the front axle, but the great angle through which the front wheels can be turned more than balances this disadvantage.

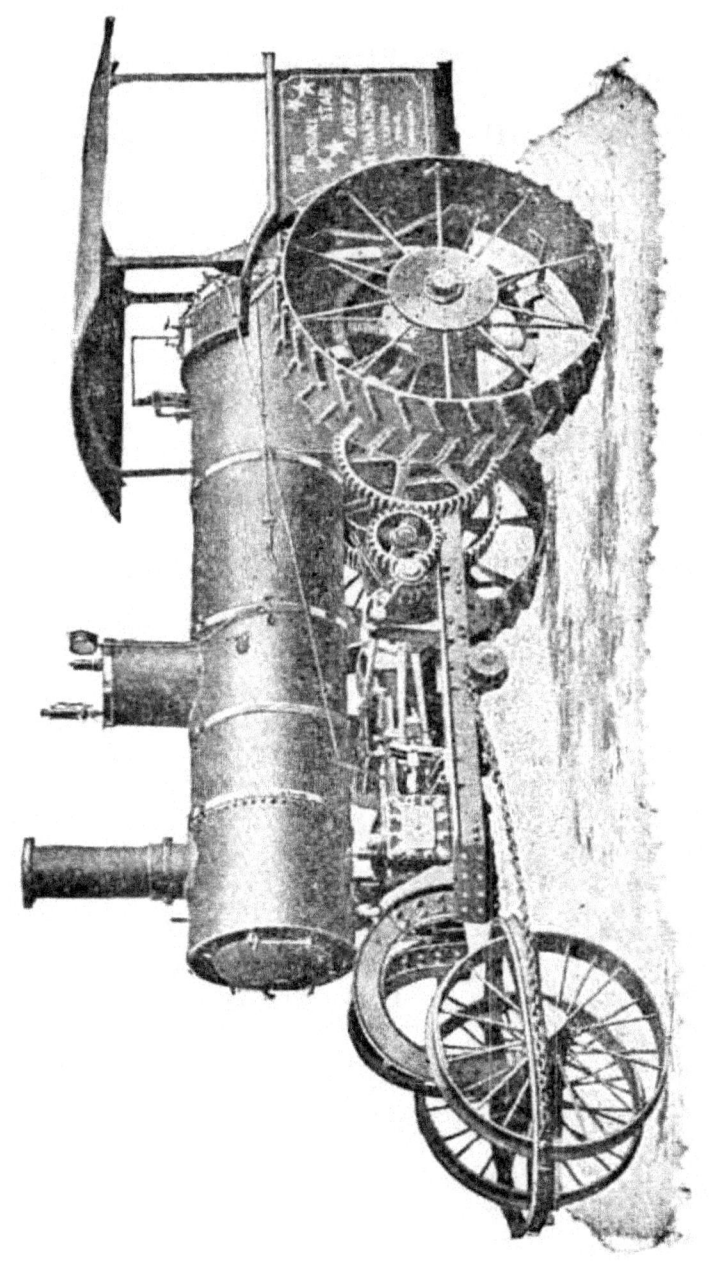

Fig. 26.—The Aultman Traction Engine.

Fig. 27.—Front View—Without Boiler—of the Aultman Traction Engine.

The boiler is of the locomotive type with the smokestack in front when coal or wood is to be used for fuel. The return tubular type with the smokestack in the rear is furnished when the fuel is to be straw. The fire-box is completely surrounded by water, and has the fire and the ash door set in a removable cast-iron frame. Practically the same fittings are furnished as in the case of the Eclipse Traction Engine.

Two independent engines with cranks at quarters and four bearings are used. I will say here that the advantage of using two engines with cranks at quarters on a traction engine over the single crank engine is quite marked. The locomotive engineers have a long time ago recognized this, and today nearly every locomotive is therefore furnished with two engines. To illustrate this the following two diagrams have been prepared. Refering to Fig. 28, which represents a diagram of a single crank engine during a quarter of a revolution of its fly-wheel, the letters A, B, C, D, and E represent positions of the crank. The figures placed opposite those letters represent the pressure of the crank-rod on the main shaft in these various positions. Supposing that the amount of pressure exerted in the position A is 100 as per diagram, then the pressure is zero when the crank is in position E or passing through the dead center. At the position B the pressure is then 75, at C 50, etc. Suppose, now, that instead of

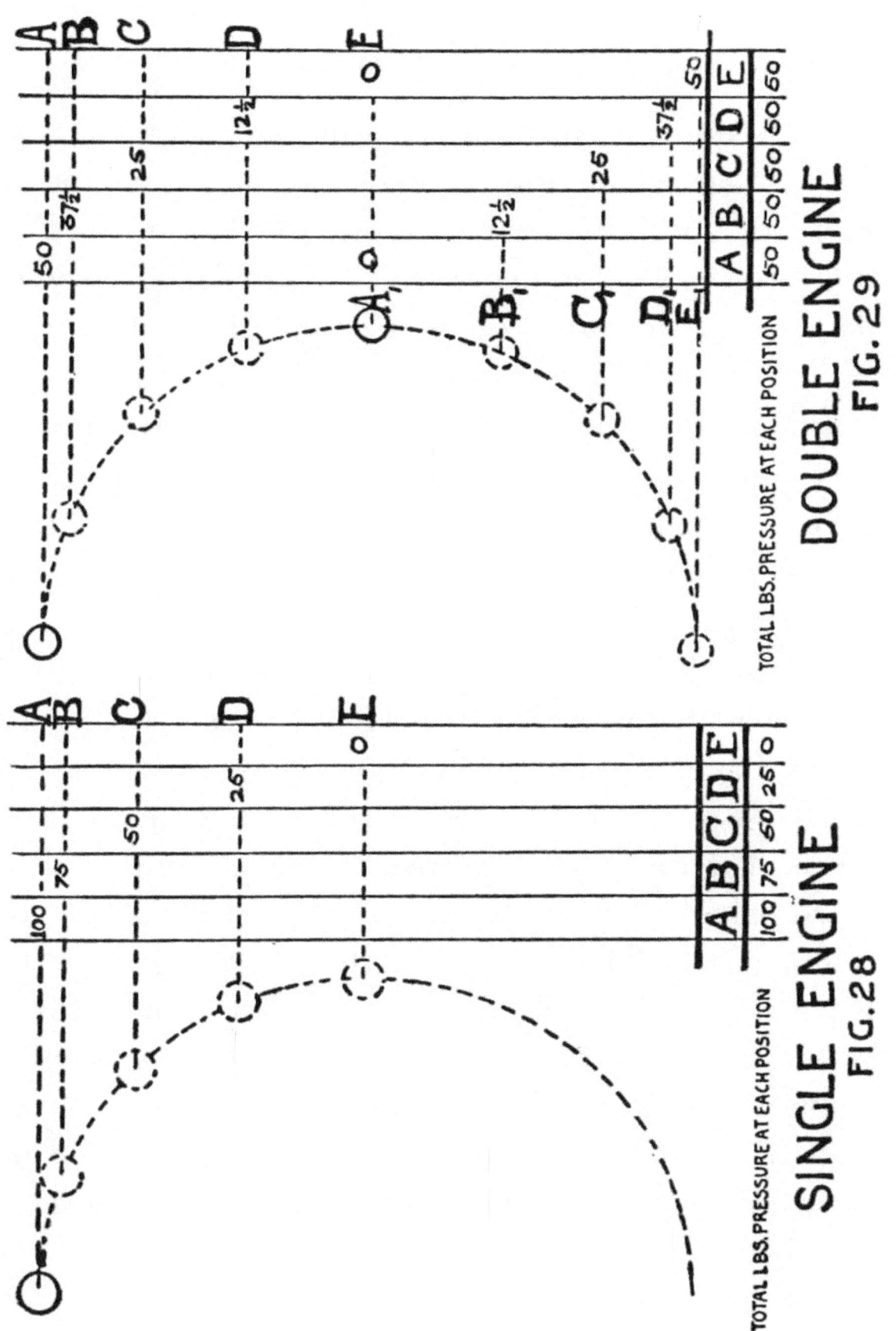

DOUBLE ENGINE
FIG. 29

SINGLE ENGINE
FIG. 28

The Traction Engine.

the single crank engine we substitute two engines with cranks at quarters having together the same amount of power as the previous single engine. Referring to Fig. 29, representing a diagram of two single crank engines with cranks at quarters during a quarter of a revolution, and allowing the letters A and A′, B and B′, etc., to represent the positions of the cranks at the same instant, then the figures set opposite those letters will also represent the pressures on the main shaft in those various positions at that instant.

As we supposed above that the total amount of power to be furnished by these two engines should be equal to the amount of power furnished by the single engine, it follows that each individual engine needs only to furnish one-half the power. By looking at the diagram we find, therefore, that the power exerted in the position A is in this case only 50, at B $37\frac{1}{2}$, at D $12\frac{1}{2}$, and at E zero for the one engine. The figures for the other engine at the same positions commencing with A′ (where it is zero) is at B′ $12\frac{1}{2}$, at D′ $37\frac{1}{2}$, and at E′ 50. As the two cranks are connected together, the total amount of pressure on the main shaft is, therefore, represented by adding together the amount of pressure which each engine exerts in corresponding positions. Doing this we find, as the diagram shows, that the total amount of pressure will be 50 in all those positions. If we now go back to the first diagram representing the single crank engine, we find that in the position A the power exerted was 100, whereas in this case we have only 50 for the double engine. At the position B the single crank engine exerts a pressure of 75, whereas the double crank engine only shows 50. At the position C both engines show the same amount of pressure, or 50. At position D the single crank engine has dropped to 25, but the double crank engine still remains 50. At position E the single crank engine exerts zero power, whereas the double crank engine still shows 50. A careful study of the above shows that, though the total amount of power developed by these two classes of engines is nearly the same, the distribution of this power during the quarter of a revolution is very materially different. The double engine exerts nearly a uniform amount of power during this period, whereas in the single engine the power varies from a maximum of nearly twice the amount of the double engine to a minimum of zero. This lack of uniformity as well as the absolute dead center have to be taken care of in the single crank engine by the fly-wheel. The double crank engine can, therefore, theoretically do without the fly-wheel. From the above we can see why the double engine with cranks at quarters is preferable for traction purposes.

The crank-shaft in this engine is of the built-up type, allowing the use of large bearings. The valves are balanced and the standard "locomotive link" reversing gear is used.

Due to the position of the engines, all their parts are well protected, easy to inspect, oil, and clean.

The height of the driving pulley on the engine shaft is convenient for handling the main belt directly from the ground without climbing, and its position is such that there is no inteference between the belt and any part of the traction engine.

The power is transmitted from the engine shaft to the traction wheels in a straight line through a train of spur gears, and the engine is provided with a set of double speed gears. The differential or compensating gear is placed inside the traction wheel. The manufacturer of this engine uses cast-steel fore gears and pinions. The traction wheels are of the built-up type with rolled steel rims and forged steel spokes and the mud cleats are of malleable cast-iron riveted to the rims.

THE NEW GIANT ENGINE.

This traction engine, which is built by the North West Thresher Co., has all its machinery mounted directly on the boiler shell. As usual with such a mounting, this engine has a very short wheel base and can therefore make very short turns. The boiler is of the return tubular flue type, with a fire-box inside the large flue. Fig. 30 shows a longitudinal sectional view of the boiler, and the end view shows the location of the return flues. The

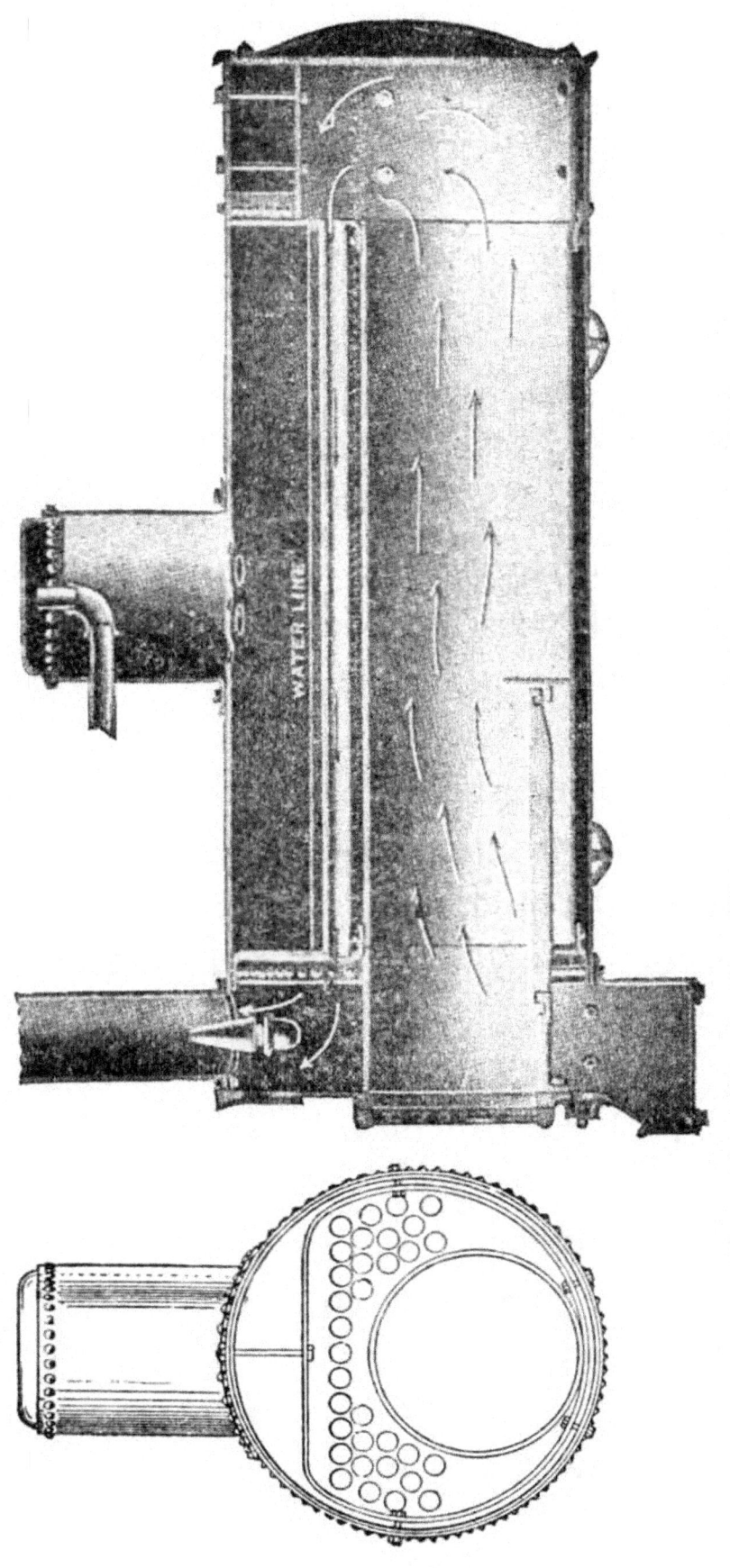

Fig. 30.—Sectional Views of the New Giant Boiler.

Shelby seamless tubes are used for the return flues, which are all located under the water-line.

The smokestack is located in the rear over the fire-door, and has the upper part hinged in order to allow the engine to pass under low bridges. A half spherical spark arrester is generally attached.

This company furnishes a Penberthy injector and a Clark independent steam-pump with their engines.

The exhaust is arranged in such a manner that it can either be directed through the feed-water heater or turned into the stack, or it may even be divided between the two so that both can be used at the same time. The exhaust nozzle is fixed and draught is regulated by directing more or less of the exhaust steam into the heater. Two safety-valves are generally used, which are set for two different pressures; that is, one of the safety-valves blows off before the other.

The water-tank is placed on top of the boiler over the front wheels, and is furnished with a steam syphon.

The engine, either simple or compound, is of the side crank type with cylinders, piston-head, cross-head, crank-disc, gearings, etc., of cast-iron. The connecting rod, shafts, and crank-pin are made of steel.

The Woolf valve gear, Gardner spring governor, sight-feed cylinder lubricator, and solid oil-cups are used.

The fly-wheel is provided with a friction clutch. All

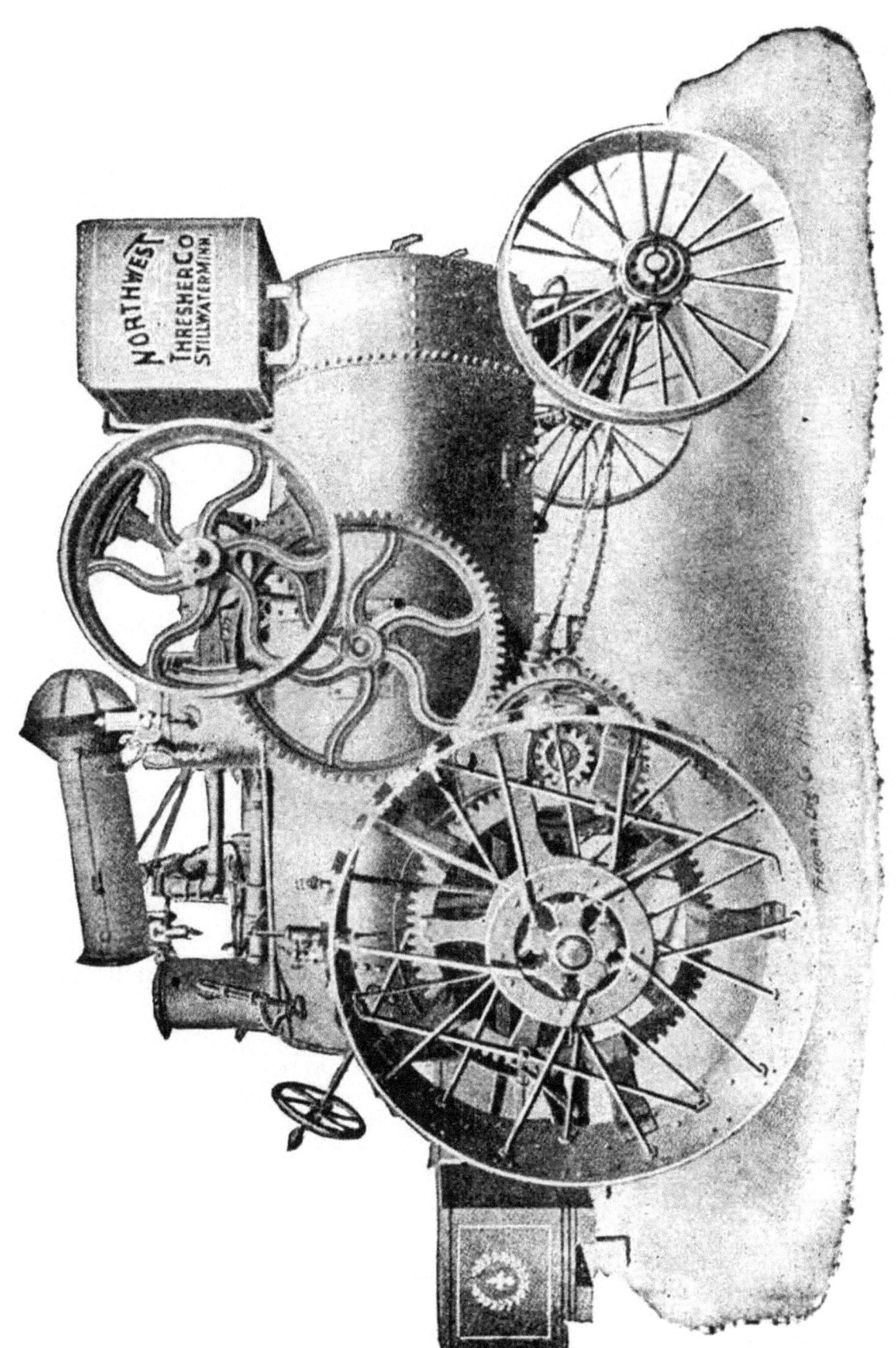

Fig. 31.—New Giant Engine, Smoke-stack Folded.

levers and handles as well as pump and injector can be operated from the engineer's platform.

The transmission of power from the engine shaft to the traction wheels is accomplished by a train of spur gears. In this train of gears is also introduced the compensating gear, which is provided with bevel steel pinions enclosed in a shell. On the outside of this shell acts a solid steel band as brake.

The traction wheels are of large diameter with broad rolled-steel tires. All spokes and braces are also made of steel. The steering chains as well as the draw-bar have strong spiral steel springs introduced between the points of attachment and themselves. Steel springs are also supporting the rear axles.

THE REEVES TRACTION ENGINE.

This engine, like the previous one, has all the machinery mounted on the boiler shell. The rear axle as well as the countershaft is supported on the rear of the boiler, and this results in a somewhat longer wheel-base than what is generally found in traction engines of this type.

The boiler is of the locomotive type of "water bottom" pattern; that is, the fire-box is surrounded by water on all sides. When engines have to be furnished for the use of straw or oil, as fuel, special arrangements of the fire-box are used. The most notable feature is the introduction in the

Fig. 32.—The Reeves Double Cylinder Traction Engine.

fire-box of a fire-brick arch made in sections and provided with a socket-joint, which permits the expansion and contraction of the arch without injury. This arch protects the crown-sheet as well as the front end of the tubes from the direct flame.

This company's traction engines are generally furnished with special rocking grates very similar in construction to the rocking grates used in some of the larger house furnaces and just as easy to handle. The controller lever is operated from the engineer's platform. The smokestack is located in front, and immediately behind it is placed the steam dome. This is made somewhat higher than what is generally found to be the practice in order to insure dry steam.

Both an injector and an independent steam pump are furnished as well as an exhaust heater.

A cylindrical water-tank is placed alongside of the fire-box and permanently piped both to injector and pump.

The engine is of the double-cylinder center-crank type with cranks at right angles. This results, as previously explained, in a machine with no dead centers. Each engine is complete and separate from the other, being only connected together by the crank-shaft. For each engine is provided an independent steam-chest, steam-valve, piston-head, cross-head, connecting-rod, etc. Only one governor of the horizontal type, one throttle-valve, and one steam-pipe is used for both engines. The steam-pipe branches to each engine below the throttle-valve. The cylinder, the frame, and the main bearing are all contained in one solid casting for each engine, and both engines are securely fastened to the same boiler saddle.

The crank-shaft is made of a steel forging in one piece and machined to size. The cross-head is provided with adjustable shoes. A reversing gear as well as an expansion controlling gear of special construction operated by a lever in the cab is used.

All fittings, such as sight-feed lubricator, steam gauge, water-glass, etc., are made of brass.

The main pulley is located on the crank-shaft and on the same side as the steering-wheel. The driving-wheels are connected by a train of spur-gears to the engine shaft. They are cast solid and provided with steel spokes cast into the hub and rims. The surface and the mud cleats have been chilled. The axle turns with the driving-wheels, and a compensating gear is introduced between it and the engine shaft.

If an engine of very high economy in regard to fuel is wanted, these engines are arranged to be cross-compounded. In this case a special valve-gear is furnished, which allows the engine to be used as a single engine with double cylinders but of larger power. When the cross-compound engine is used in the latter way, the economy is, of course,

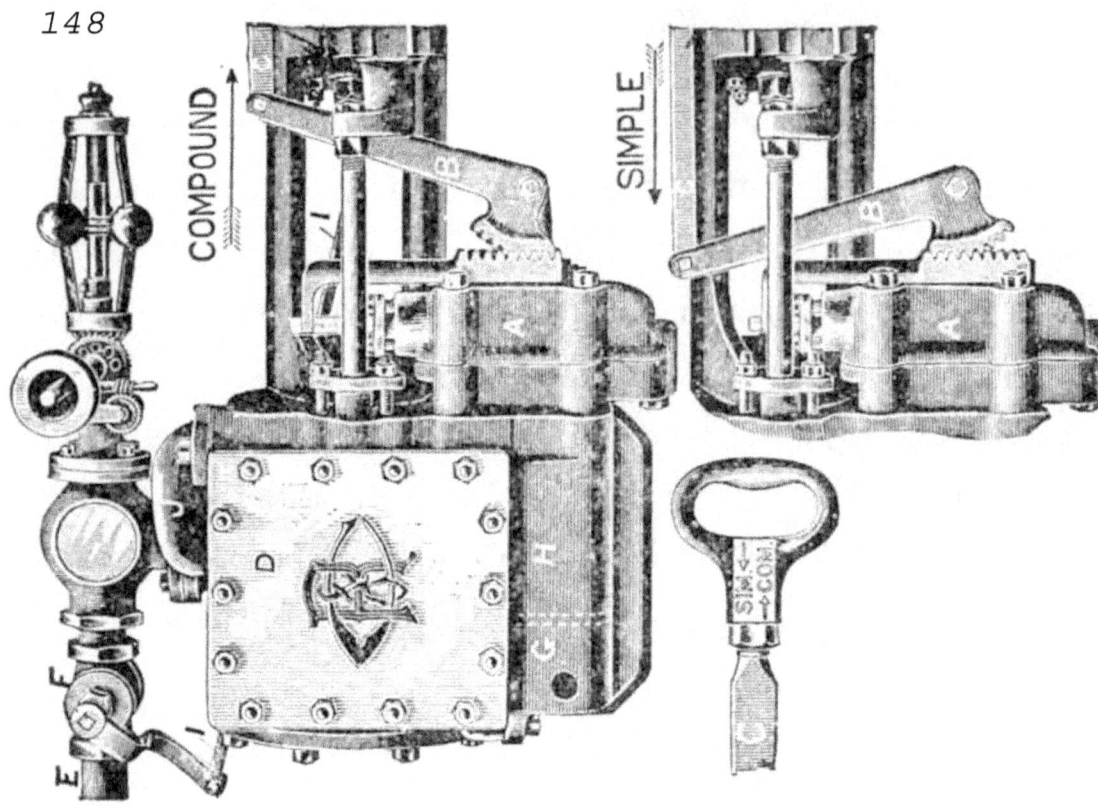

Fig. 34.—The Reeves Governor and Valve Gear for Changing from a Simple to a Compound Engine.

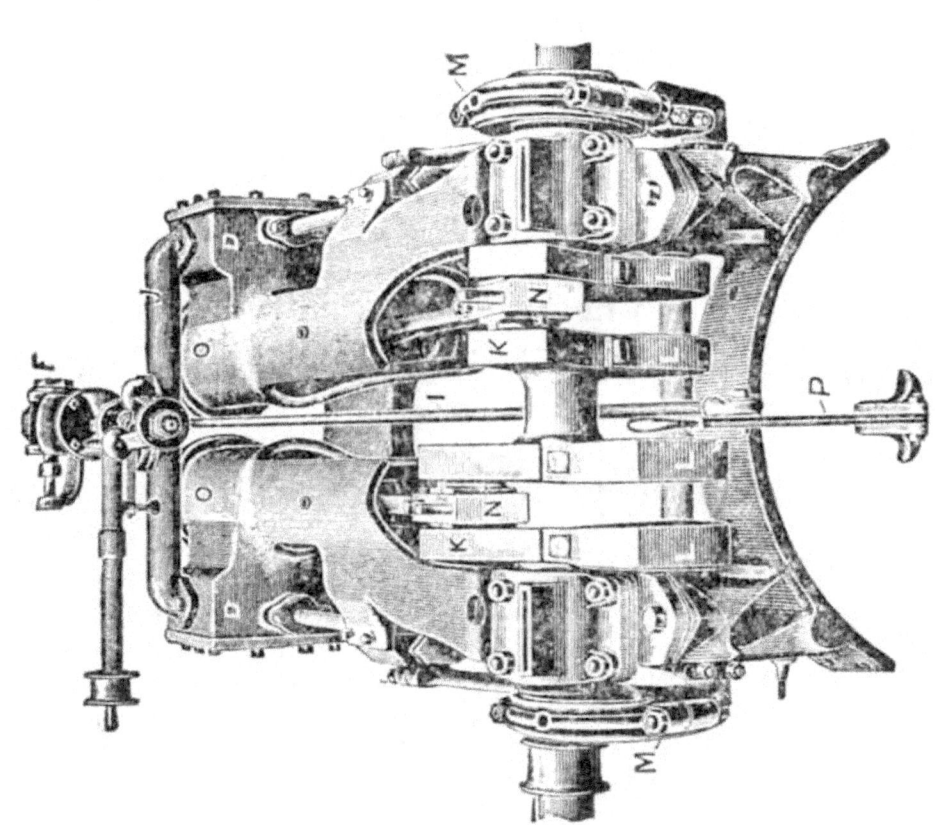

Fig. 33.—The Reeves Double Engines.

no greater than that of a single engine. Fig. 34 shows a cut of this valve gear.

THE COLEAN TRACTION ENGINE.

This is another engine of the same type as the two previous ones just described, having all its machinery mounted directly on the boiler shell. It has the same advantage as most engines of this class have; that is, its wheel base is very short.

The boiler, of the locomotive type, is built of steel with double-riveted seams, and has water-jacketed fire-box and rocking grates. The latter are controlled from the engineer's platform by means of a lever.

The smokestack is in front and the steam dome has a relatively large steam space. The usual fittings are supplied. The engines, of the Corliss self-contained type with double cylinder and crank at quarters, are mounted side by side on the top of the boiler. The steam-chest and the cylinder are cast in one piece. The crank-shaft is made of a solid steel-forging machined to proper dimensions. The locomotive reversing links, horizontal governor, sight-feed lubricator, etc., are all standard. The fly-wheel or pulley is provided with a friction clutch so arranged that it can be converted into a solid pulley by the shifting of a pin and the tightening of a set screw, which operation practically makes a solid connection between the traction pinion, the fly-wheel, and the engine shaft.

Fig. 35.—The Colean Traction Engine.

Different Traction Engines.

The water-tank is placed in front on top of the boiler and surrounds the smokestack. The driving-wheels have steel rims and spokes and are connected to the engine through spur-gearing. In this train of gears is inserted the compensating gear, which has bevel steel pinions.

BUFFALO PITTS CO. TRACTION ENGINE.

This company mounts also the machinery directly on the boiler shell and secures, therefore, the advantage of a short wheel base.

The boiler is of the locomotive type, with Shelby's cold drawn seamless tubes aud full water bottom. On the water bottom the fire is dumped instead of dumping it directly on the ground, thereby preventing the starting of fiers. The boiler can be arranged for burning either coal, wood, or straw. When only straw is to be used for fuel, the return tubular flue type of boiler is generally used. Whatever the type of boiler may be, a fire-brick arch is put in the fire-box as well as special straw-burning grate bars.

All longitudinal seams are double-riveted. The steam space has been made relatively large and so has the dome in order to secure as dry steam as possible.

Both a pump and an injector are furnished, and have been so arranged that either can be used for filling the tank as well as taking water for the boiler from the tank

The tank, holding about 100 gallons of water is of steel and located on the front of the boiler.

FIG. 36.—THE BUFFALO PITTS TRACTION ENGINE.

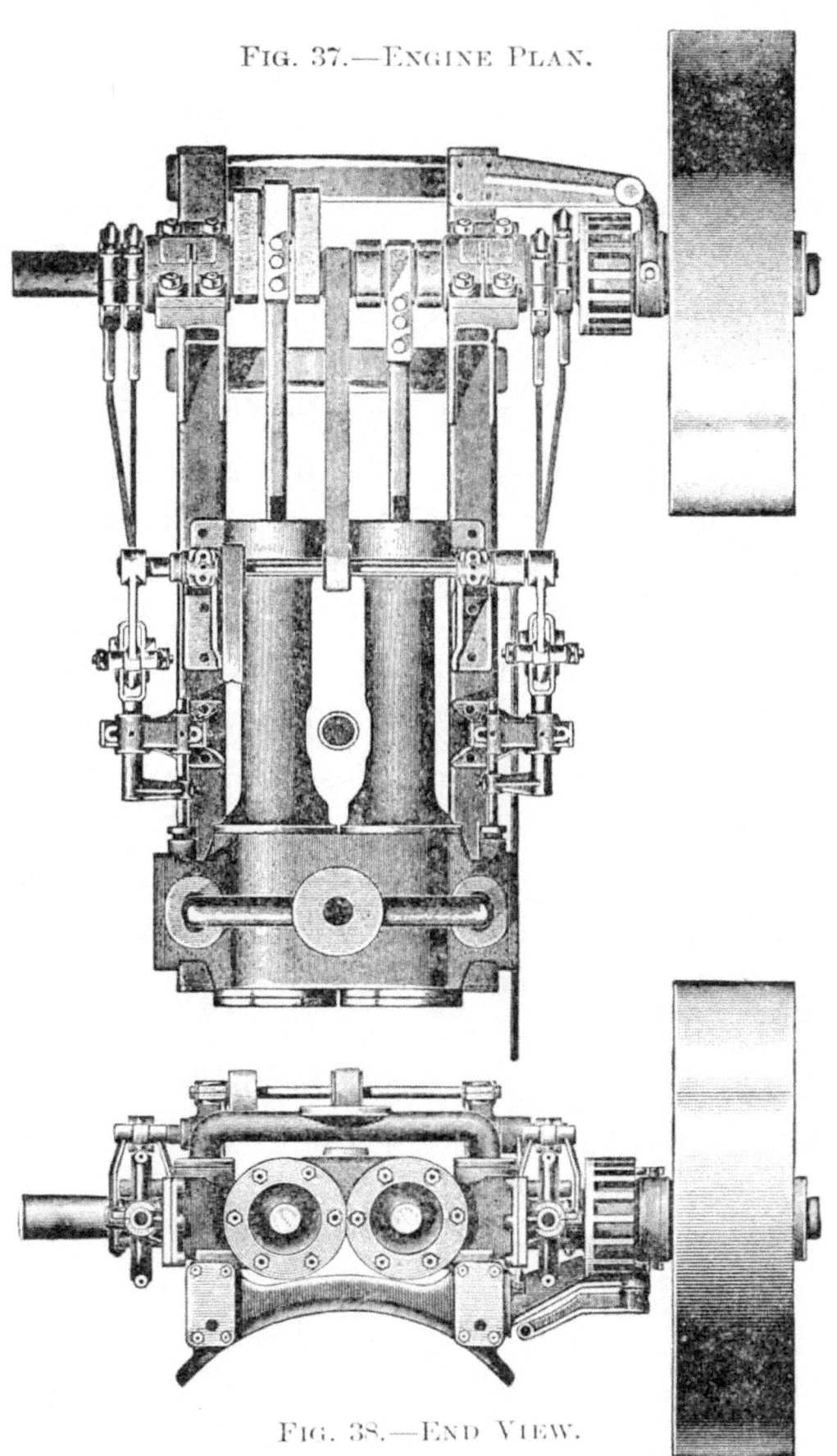

Fig. 37.—Engine Plan.

Fig. 38.—End View.

Two independent engines of the center-crank self-contained type are mounted side by side. Each has its own steam-chest, piston-rod, cross-head, connecting-rod, etc. Figs. 37 and 38 give you a view of the engine. The cylinders are fastened together on the inside of the frame and have the steam-chests located on the outside. The cylinders and the guides, cast in one piece, are bored out together and therefore always in line. Due to the construction the moving parts of the engine are always well protected, and can be very easily inspected, oiled, and cleaned. The governor is of the vertical type, and is driven by a small belt from a little counter-shaft located on the front part of the guides. (See Fig. 36.) It is mounted on the pipe flange just before the steam-pipe branches to the cylinders.

The locomotive reversing links are used. The crankshaft is made of a steel forging with cranks at quarters and is provided with self-oiling attachments. A train of spur-gears transmits the power from the engine to the driving-wheels. In this train is introduced the compensating gear containing three bevel pinions. The maker claims that this number is preferable, as it prevents the rocking motion and therefore reduces the wear.

The traction wheels are of the built-up type with broad rolled steel rims and flat steel spokes riveted to the rim and the hub. Large mud-cleats are also riveted on the outside of the rim of the wheels. The fly-wheel has been provided with a friction clutch.

FIG. 39.—THE HUBER TRACTION ENGINE.

THE HUBER TRACTION ENGINE.

This engine belongs to the same class as the previous ones which are using their boilers as frames for supporting the machinery. The wheel base is for this reason very short. This company uses the return tubular type of boiler with the fire-box inside the large central tube and with the smokestack in the rear. It has a special superheating device (see

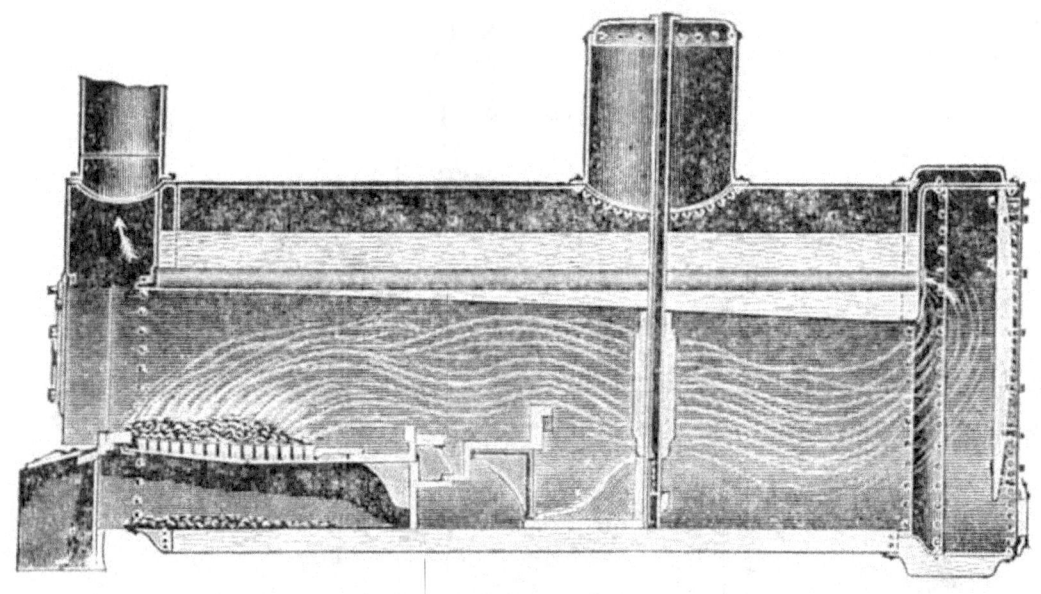

Fig. 40.—The "Huber" Boiler.

Fig. 40) consisting of a double tube, the outer one reaching from the steam dome down into the large fire-tube, and the inner one, constituting the steam delivery pipe, starting near the bottom of the outer one, carries the superheated steam through the dome to the engine. The combustion chamber in the front is provided with a hinged

Different Traction Engines.

door on which the water-tank is mounted. By swinging back this door all the tubes are exposed and can be easily cleaned and inspected.

The smokestack is constructed of several layers of sheet-iron with air-spaces between, which construction leaves the outside surface of the stack relatively cool.

A variable exhaust is secured by having the exhaust nozzle divided by a central partition in two compartments

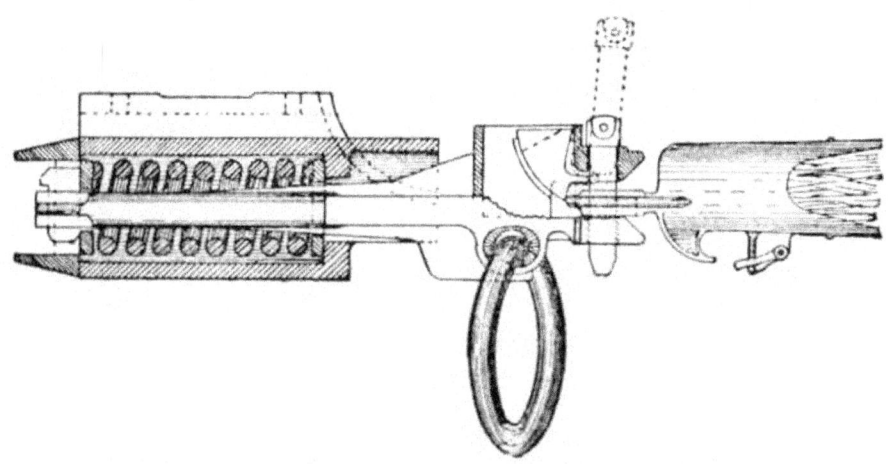

Fig. 41.—The Spring Draw Bar.

and regulating the amount of steam blowing through one of them by a valve.

A cross-head pump, a syphon (see Fig. 16), and an injector take care of the water-supply.

The engine is of the side-crank single-cylinder type, with cylinder, guide, and crank-shaft bearings all in one casting. It has a belt-driven horizontal governor and the Huber reversing gear. The connection between the crank-shaft and the driving-wheel consists of a train of spur-gears. In this train of gears is also included a compensating gear which

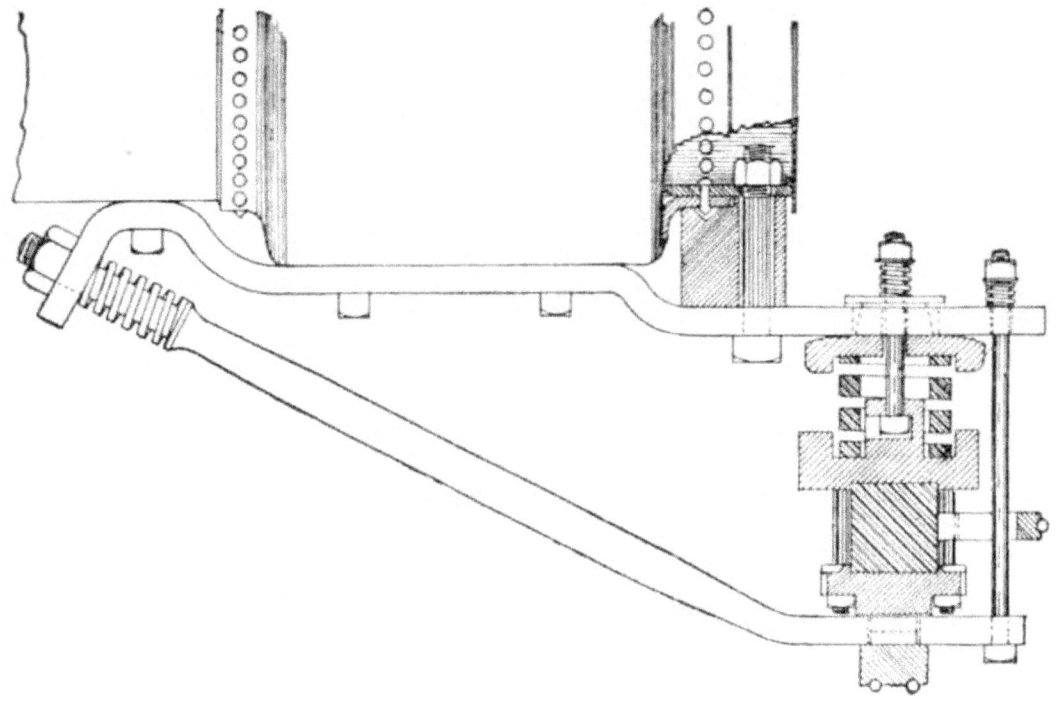

Fig. 42.—Spring Mounting.

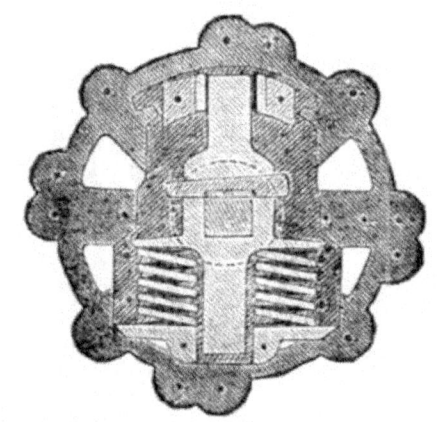

Fig. 43.—Section of Hub.

is constructed with spur-gears for pinions. This compensating gear is enclosed in a casing and covered by a steel plate, making it perfectly dust-free. Between the last gear and the driving-wheels four spiral springs are introduced. Springs are also used between the drivers and the axle (see Fig. 43), between the front axle and the boiler (see Fig. 42), and between the draw-bar and the engine (see Fig. 41).

The driving-wheels are of the built-up type with flat drop forged-steel spokes and rolled steel rims. The axle is square, does not revolve, and passes under the boiler from one wheel to the other.

THE PORT HURON TRACTION ENGINE.

This engine is built by the Port Huron Engine and Thresher Co., and belongs to the same class as the one just described; that is, it has all its machinery mounted on the boiler and also a short wheel base.

The boiler belongs to the locomotive type with smoke-stack in front. The fire-box has a circular bottom and is water-jacketed all around. The waist seam, connecting the fire-box to the boiler, is double riveted and the tubes are lapwelded with a copper ferrule for tightener between the tubes and the tube sheet. The furnace door and its frame are of cast-iron and bolted to the fire-box, allowing it to be removed for repair. The smoke-box is relatively long.

A cross-head pump with condensing heater and an injector furnish the boiler with water.

Fig. 44.—Port Huron Traction Engine.

Different Traction Engines.

The engine is of the side crank type with a girder-like frame and bored guides. Instead of the usual slide valve of the D type this company makes use of four independent valves of the poppet type, resulting in a square cut-off very similar to the cut-off of the Corliss engine. The valve (see Fig. 45) consists of a hollow cylindrical body with circular discs on each end fitting the valve seats. The disc on the boiler side is of a little larger area than the one on the cylinder side, thus keeping the valve seated by the excess of steam pressure due to this difference in area and assisted by a light spring. The valves are operated by cams on a cylindrical rotating shaft deriving its motion from the crank-shaft by means of bevel gears. (See Fig. 46.) This shaft opens and closes the valves at the right time by the cams operating on the stems of the valve. The reversing as well as the regulating of the engine is accomplished by shifting this shaft along its own axle by means of a lever in the cab.

A key-way in the bevel gears and a key in the shaft keeps them always in the right position relatively to each other. This valve gear allows the valves to be kept wide open as long as wanted and then cuts off very quickly, preventing all wire drawing of the steam.

A train of spur-gears connects the crank-shaft with the driver. Each driving-wheel axle is supported on a bracket fastened on the side of the fire-box. The bracket is con-

Fig. 45.

Fig. 47.

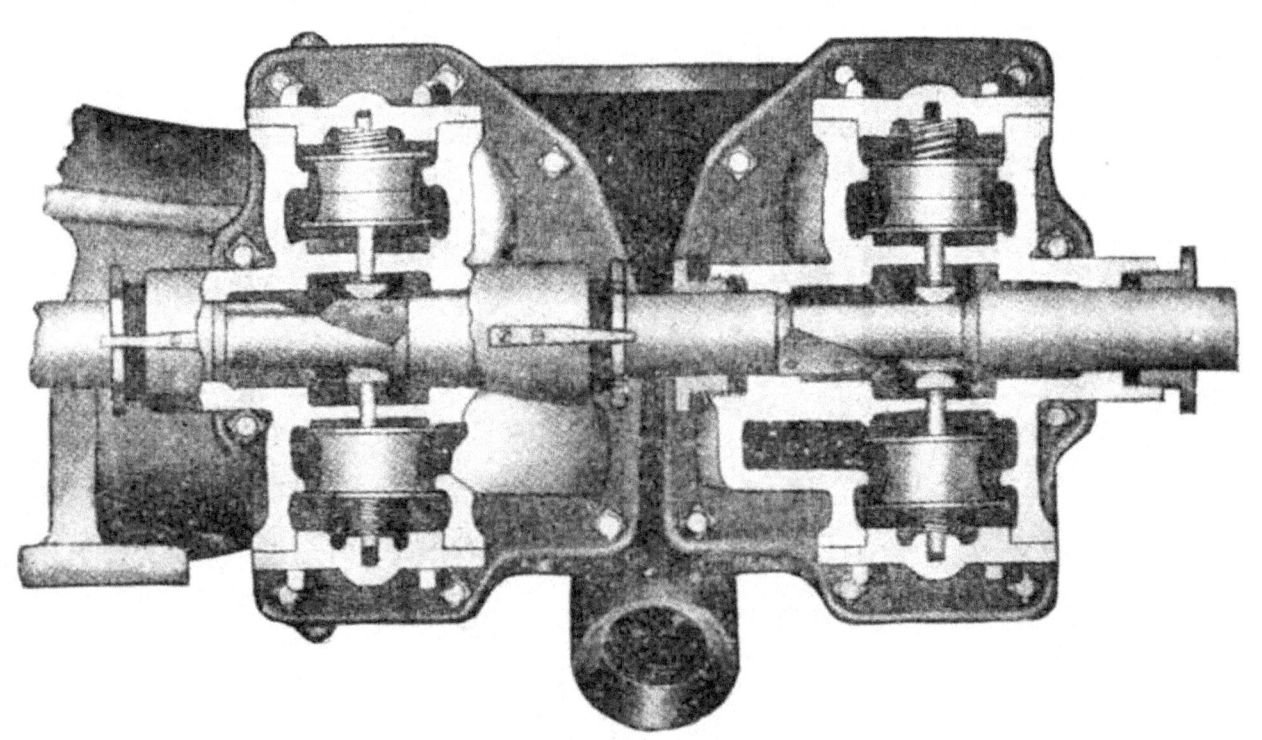

Fig. 46.

nected with the bracket on the opposite side of the fire-box by a tie rod under the bottom (see Fig. 47). The driving-wheels are solid, no springs being interposed between them and the boiler. The same is the case with the front wheels. In fact, absence of springs is a notable feature about this engine.

PART NINTH.

SOME THINGS TO KNOW.

Q. How do you find the circumference of a circle?

A. Multiply the diameter by 3 or, more correctly, by $3\frac{1}{7}$.

Q. How do you find the diameter of a circle when you know the circumference?

A. Multiply by 318 and divide by 1000.

Q. How do you find the area of a circle?

A. Multiply the diameter by itself and multiply the result by 785 and divide by 1000.

Q. What is the weight of a gallon of water?

A. Eight and one-third pounds.

Q. How many gallons are contained in one cubic foot of water?

A. Seven and one-half gallons.

Some Things to Know.

Q. What is combustion?

A. A chemical combination of oxygen and carbon.

Q. What is fire?

A. Fire is the rapid combustion or consuming of organic matter.

Q. What is water?

A. Water is a compound of oxygen and hydrogen. In weight, $88\frac{9}{10}$ parts oxygen to $11\frac{1}{10}$ hydrogen. It has its maximum density at 39 degrees Fahr., changes to steam at 212 degrees, and to ice at 32 degrees.

Q. What is smoke?

A. It is unconsumed carbon, finely divided, escaping into open air.

Q. Is excessive smoke a waste of fuel?

A. Yes.

Q. How will you prevent it?

A. Keep a thin fire, and admit cold air sufficient to insure perfect combustion.

Q. What is low water as applied to a boiler?

A. It is when the water is insufficient to cover all parts exposed to the flames.

Q. What is the first thing to do on discovering that you have low water?

A. Pull out the fire.

Q. Would it be safe to open the safety-valve at such time?

A. No.

Q. Why not?

A. It would relieve the pressure on the water and a large portion of the super-heated water would flash into steam and cause an explosion.

Q. Why do boilers sometimes explode just on the point of starting the engine?

A. Because starting the engine has the same effect as opening the safety-valve.

Q. Are there any circumstances under which an engineer is justified in allowing the water to get low?

A. No.

Q. Why do they sometimes do it?

A. From carelessness or ignorance.

Q. May not an engineer be deceived in the gauge of water?

A. Yes.

Q. Is he to be blamed under such circumstances?

A. Yes.

Q. Why?

A. Because if he is deceived by it, it shows he has neglected something.

Q. What is meant by "priming"?

A. It is the passing of water in visible quantities into the cylinder with the steam.

Q. What would you consider the first duty of an

engineer on discovering that the water was foaming or priming?

A. Open the cylinder cocks at once, and throttle the steam.

Q. Why would you do this?

A. Open the cocks to enable the water to escape, and throttle the steam so that the water would settle.

Q. Is foaming the same as priming?

A. Yes and no.

Q. How do you make that out?

A. A boiler may foam without priming, but it can't prime without first foaming.

Q. Where will you first discover that the water is foaming?

A. It will appear in the glass gauge, the glass will have a milky appearance, and the water will seem to be running down from the top. There will be a snapping or cracking in the cylinder as quick as priming begins.

Q. What causes a boiler to foam?

A. There are a number of causes. It may come from faulty construction of boiler; it may have insufficient steam room. It may be, and usually is, from the use of bad water, muddy or stagnant water, or water containing any soapy substance.

Q. What would you do after being bothered in this way?

A. Clean out the boiler and get better water if possible.

Q. How would you manage your pumps while the water was foaming?

A. Keep them running full.

Q. Why?

A. In order to make up for the extra amount of water going out with the steam.

Q. What is "cushion?"

A. A cushion is steam retained or admitted in front of the piston head at the finish of stroke, or when the engine is on "center."

Q. What is it for?

A. It helps to overcome the "inertia" and momentum of the reciprocating parts of the engine, and enables the engine to pass the center without a jar.

Q. How would you increase the cushion in an engine?

A. By increasing the lead.

Q. What is lead?

A. It is the amount of opening the port shows on steam end of cylinder when the engine is on dead center.

Q. Is there any rule for giving an engine the proper lead?

A. No.

Q. Why not?

A. Owing to their variation in construction, speed, etc.

Some Things to Know.

Q. What would you consider the proper amount of lead, generally?

A. From $\frac{1}{32}$ to $\frac{1}{16}$.

Q. What is "lap?"

A. It is the distance the valve overlaps the steam ports when in mid position.

Q. What is lap for?

A. In order that the steam may be worked expansively.

Q. When does expansion occur in a cylinder?

A. During the time between which the port closes and the point at which the exhaust opens.

Q. What would be the effect on an engine if the exhaust opened too soon?

A. It would greatly lessen the power of the engine.

Q. What effect would too much lead have?

A. It would also weaken the engine, as the steam would enter before the piston had reached the end of the stroke, and would tend to prevent its passing the center.

Q. What is the stroke of an engine?

A. It is the distance the piston travels in the cylinder.

Q. How do you find the speed of a piston a minute?

A. Double the stroke, and multiply it by the number of revolutions a minute. Thus, an engine with a twelve-inch stroke would travel twenty-four inches, or two feet, at a revolution. If it made 200 revolutions a minute, the travel of piston would be 400 feet a minute.

Q. What is considered a horse-power as applied to an engine?

A. It is a power sufficient to lift 33,000 pounds one foot high in one minute.

Q. What is the indicated horse-power of an engine?

A. It is the actual work done by the steam in the cylinder as shown by an indicator.

Q. What is the actual horse-power?

A. It is the power actually given off by the driving belt or pulley.

Q. How would you find the horse-power of an engine?

A. Multiply the area of the piston by the average pressure less five; multiply this product by the number of feet the piston travels in a minute; divide the product by 33,000; the result will be the horse-power of the engine.

Q. How will you find the area of piston?

A. Square the diameter of piston, and multiply it by .7854.

Q. What do you mean by squaring the diameter?

A. Multiplying it by itself. If a cylinder is six inches in diameter, 36 multiplied by .7854 gives the area in square inches.

Q. What do you mean by average pressure?

A. If the pressure on boiler is sixty pounds, and the engine is cutting off at $\frac{1}{2}$ stroke, the average pressure for the full stroke would be fifty pounds.

Some Things to Know.

Q. Why do you say "less five pounds"?
A. To allow for friction and condensation.

Q. What is the power of a 7 x 10 engine, running 200 revolutions, cutting off at ½ stroke, with 60 pounds steam?
A. 7 x 7 = 49 x .7854 = 38.48. The average pressure of 60 pounds would be 50 pounds less 5 = 45 pounds; 38.48 x 45 = 1731.80 x .333⅓ (the number of feet the piston travels a minute) = 577,269.00 ÷ by 33,000 = 17½ horse-power.

Q. What is a high-pressure engine?
A. It is an engine using steam at a high pressure and exhausting into the open air.

Q. What is a low-pressure engine?
A. It is one using steam at a low pressure and exhausting into a condenser, producing a vacuum, the piston being under steam pressure on one side and vacuum on the other.

Q. What class of engines are farm engines?
A. They are high-pressure.

Q. Why?
A. They are less complicated and less expensive.

Q. What is the most economical pressure to carry on a high-pressure engine?
A. From 90 to 110 pounds.

Q Why is high pressure more economical than low pre   ure?

A. Because the loss is greater in low pressure owing to the atmospheric pressure. With forty-five pounds steam the pressure from the atmosphere is fifteen pounds, or $\frac{1}{3}$, leaving only thirty pounds of effective power; while with ninety pounds the atmospheric pressure is only $\frac{1}{6}$ of the boiler pressure.

Q. Does it require any more fuel to do the work if we carry 100 pounds than it does to carry sixty pounds?

A. It does not require quite so much.

Q. If that is the case, why not increase the pressure beyond this and save more fuel?

A. Because we would soon pass the point of safety in a boiler, and the result would be the loss of life and property.

Q. What do you consider a safe working pressure on a boiler?

A. That depends entirely on its diameter. While a boiler of thirty inches in diameter, $\frac{3}{8}$ inch iron, would carry 140 pounds, a boiler of the same thickness eighty inches in diameter would have a safe working pressure of only fifty pounds, which shows that the safe working pressure decreases very rapidly as we increase the diameter of boiler. This is the safe working pressure for single-riveted boilers of this diameter. To find the safe working pressure of a double-riveted boiler of same diameter, multiply the safe pressure of the single-riveted by seventy, and divide by fifty-six; it will give the safe pressure of a double-riveted boiler.

Q. Why is a steel boiler superior to an iron boiler?

A. Because it is much lighter and stronger.

Q. Does boiler plate become stronger or weaker as it becomes heated?

A. It becomes tougher or stronger as it is heated, until it reaches a temperature of 550 degrees, when it rapidly decreases its power of resistance as it is heated beyond this temperature.

Q. How do you account for this?

A. Because after you pass the maximum temperature of 550 degrees, the more you raise the temperature, the nearer you approach its fusing-point when its tenacity, or resisting power, is nothing.

Q. What is the degree of heat necessary to fuse iron?

A. Nearly 4000 degrees.

Q. What class of boilers are generally used in a threshing engine?

A. The flue boiler and the tubular boiler.

Q. About what amount of heating and grate surface is required per horse-power in a flue boiler?

A. About fifteen square feet of heating surface and $\frac{3}{4}$ of a square foot of grate surface.

Q. What would you consider a fair evaporation in a flue boiler?

A. Six pounds of water to one pound of coal.

Q. How do these dimensions compare in a tubular boiler?

A. A tubular boiler will require ¼ less grate surface, and will evaporate about eight pounds of water to one pound of coal.

Q. Which do you consider the most available?

A. The tubular boiler.

Q. Why?

A. It is more economical and is less liable to "collapse."

Q. What do you mean by "collapse"?

A. It is a crushing in of a flue by external pressure.

Q. Is a tube of large diameter more liable to collapse than one of smaller diameter?

A. Yes.

Q. Why?

A. Because its power of resistance is much less than that of a tube of small diameter.

Q. Is the pressure on the shell of a boiler the same as on the tubes?

A. No.

Q. What is the difference?

A. The shell of boiler has a tearing or internal pressure, while the tubes have a crushing or external pressure.

Q. What causes an explosion?

Some Things to Know.

A. An explosion occurs generally from low water, allowing the iron to become overheated and thereby weakened and unable to withstand the pressure.

Q. What is a "burst?"

A. It is that which occurs when, through any defect, the water and steam are allowed to escape freely without further injury to boiler.

Q. What is the best way to prevent an explosion or burst?

A. (1) Never go beyond a safe working pressure. (2) Keep the boiler clean and in good repair. (3) Keep the safety-valves in good shape and the water at its proper height.

Q. What is the first thing to do on going to your engine in the morning?

A. See that the water is at its proper level.

Q. What is the proper level?

A. Up to the second gauge.

Q. When should you test or try the pop-valve?

A. As soon as there are a few pounds of steam.

Q. How would you start your engine after it had been standing over night?

A. Slowly.

Q. Why?

A. In order to allow the cylinder to become hot, and

that the water or condensed steam may escape without injury to the cylinder.

Q. What is the last thing to do at night?

A. See that there is plenty water in boiler, and, if the weather is cold, drain all pipes.

Q. With a new engine and boiler, what will you do to start it?

A. (1) Examine all parts and oil all bearings. Turn the engine by hand to see that it moves freely. (2) Fill the boiler up to the second gauge cock by means of the force-pump. (3) Start the fire slowly, and when steam-gauge shows five pounds, turn on the blower to increase the burning. (4) Open cylinder cocks. (5) Open throttle-valve gradually and let the engine turn over. If everything appears to go all right, open the throttle in full. (6) When only steam comes out the cylinder cocks, close them. (7) Examine bearings. (8) Start pump or injector to feeding just fast enough to supply water as fast as it goes out of the boiler into the engine.

Q. What would you do in cold weather if you thought that your pumps, boiler connections, or water-pipes were liable to be frozen?

A. I would open all drip- and discharge-cocks, and allow all the water to run out of them when I stop work at night. In the morning I would examine all the steam- and water-connections before starting the fire.

Some Things to Know.

Q. What would you do if you must stop your engine when the steam is blowing off at the safety-valve?

A. I would immediately start the pump or injector and cover the fire with fresh coal.

Q. What is the use of a fusible plug?

A. It is intended for a safety device, in case the water gets too low in the boiler.

Q. How does the fusible plug act?

A. When the water in the boiler gets so low as to uncover the plug, the fire in the fire-box will melt it and the steam which will flow through the opening into the fire-box will extinguish the fire.

Q. Where is the fusible plug generally placed in a boiler?

A. In the crown-sheet.

Q. How is the fusible plug made?

A. It is made of a short piece of brass tubing threaded on the outside and having a hexagonal head on one end. This tube is then filled with a metal compound melting at a relatively low temperature.

PART TENTH.

INTERNAL COMBUSTION ENGINES.

GENERAL PRINCIPLES.

"Internal combustion engines" is the general name for the Gas Engine, the Gasoline Engine, the Oil Engine, the Alcohol Engine and others. As you can understand from the name, they use either gas, gasoline, coal oil or alcohol for fuel.

Why are they called internal combustion engines? Well, the reason is that they burn their fuel inside their cylinders. We have seen, in the first pages of this book, that the steam engine has to be supplied with steam, that the steam has to be generated from water, and that we have to burn fuel under or inside the boiler to make the steam. We say that all engines which use fuel in this way belong to the steam engine class. In a similar manner we say that all machines burning their fuel direct in their cylinders belong to the "internal combustion engine" class.

Internal Combustion Engines.

Now, it is easy to see that different principles must be considered when dealing with the one class than with the other. We are going to tell you how to run a gas engine and a gasoline engine, but before we do that we must tell you something about the principles which makes it possible for this class of engines to run. Not because this knowledge will make you run your engine better, but because, if for any reason your engine will not run, you must be able to make it run.

In order to do this you must know these things, and then you must use your brains. There is no fire-door on this class of engines and no steam-gauge, so you cannot open the one or look at the other for information. When your engine don't behave right you must be able to reason out what can be the trouble from your knowledge of what ought to take place. Every year you will meet with more and more gas and gasoline engines, and therefore I want to try to make you thoroughly understand this subject.

Now turn to page 194 and look at figure 48. This will give you a very good idea of a stationary gasoline engine of a good, modern design. It looks very much like a steam engine without the boiler.

Before we talk more of internal combustion engines we will say a few words about the steam engines. As you very well know, the steam engine cylinder is generally closed at both ends and steam enters first at one end and then at the other. This makes the ordinary steam-engine double-acting. Or, in other words, for each revolution of the crank the fly-wheel receives two power impulses. There are other steam engines, like the Westinghouse, the Brotherhood, and others, which have only one end of the cylinder closed and therefore can only take steam on one end. Those engines are called "single-acting," and they get only one power impulse for each revolution of their fly-wheel. Of course, if you put two such cylinders on one engine, the fly-wheel will get two power impulses for each revolution, but the engine is nevertheless only a single-acting engine. Now, bearing in mind what we have just said, you will understand what I mean when I make the statement that most of the internal combustion engines are single-acting in their construction.

You notice, I made the addition "in their construction," and the reason is, as you will see presently, that though the engine is single-acting the fly-wheel will, however, only receive one impulse for every *two* revolutions, or just one-half as many as in the case of the single-acting steam engine.

You have often seen a steam engine shut down. When the steam is shut off, the engine makes only a few revolutions and then stops. Now watch a gasoline engine, and you will notice that it runs very much longer. If you take another look at our figure 48 you will notice that it

Internal Combustion Engines.

has two fly-wheels and that they look pretty heavy. A double-acting steam engine does not need such heavy fly-wheels because it gets two impulses for each revolution of the crank, as we saw above. Even the single-acting steam engine does not need as heavy ones, because it gets one impulse for each turn of the wheels, but the gasoline engine cannot get along without them, because it gets only one impulse for every two revolutions of the fly-wheels.

From this you can conclude that the weight of the fly-wheel gets heavier for the same regulation and speed, as the number of power impulses the wheels receive per revolution is diminished. You will often hear it said that a gasoline engine does not regulate as well as a steam engine, and that is correct; but if you put on heavy enough fly-wheels you can make it regulate just as well. Of course, an internal combustion engine, which would receive more impulses per revolution of fly-wheel, would need a smaller fly-wheel for the same amount of regulation.

You remember that the internal combustion engine burns its fuel inside the cylinder. That is, a certain amount of fuel and air (fuel cannot burn without air) intimately mixed with each other are introduced into the cylinder. This mixture is compressed by the action of the engine and at the proper time ignited inside the cylinder. The burning gases expand and push the piston forward, thus furnishing power. It is very important that the mixture gets the proper amount of compression, because this not only brings the air and the fuel into closer contact with each other, but it results also in warming up the mixture, which under those conditions is easier ignited.

We will now see how this ignition is accomplished, and we will mention the various devices in the same order as they were brought out.

Open flame ignition was used in combination with a slide-valve. This kind of ignition is only found on very old engines and is not used any more.

Tube ignition is still used on some makes of engines, but is gradually disappearing. It consists of a hollow tube of nickel, steel, iron, or even porcelain, kept at a temperature varying from red to yellowish-red. The heat for this purpose is generally furnished by a "Bunsen" burner, "Gasoline torch," or some similar device playing on the outside of the tube. The compressed charge coming in contact with the hot walls of this tube causes the explosion. This kind of ignition is subject to various drawbacks. The most important of these is the impossibility of fixing the time when the ignition is to take place, as this depends, among many other conditions, on the exact temperature of the tube, the proportion of fuel and air in the charge, and the temperature of the charge itself.

Internal Combustion Engines.

Electric Ignition. It is called electric because it fires the compressed charge by an electric spark, which is made to jump between two points inside the cylinder. This is the way a charge is ignited in a modern internal combustion engine.

Before saying anything more, I will tell you that the point from which the spark jumps and the point to which it jumps are both called "electrodes." This is the regular name for them all over the country, and you may just as well know it now as later.

There is a great number of electrical igniters. Nearly every maker of an engine uses a different kind. They belong, however, all to one or the other of the following two classes. Namely:

1. Igniter with stationary electrodes.
2. Igniter with movable electrodes.

The electric igniter with either kind of electrodes must be supplied with a current of electricity. This is generally furnished by a battery.

We suppose that you are well acquainted with such an apparatus. Probably you have an electric battery in your own house which rings your door-bell or your telephone bell. There are a great number of makes of batteries, and each maker of an engine has found out which particular one is best suited for his engine, and that is the kind you must use.

Some makers use a small dynamo or "sparker," bolted right on to the engine, and driven by belt or gears from the engine. In that case you do not need a battery.

In either case you must furnish electricity of the proper kind suitable for your electrodes if you want your igniter to work. Generally you will find that batteries are used, and if you have stationary electrodes you will find that an apparatus called an induction coil is also supplied. With this kind, the current from the battery goes into what is called the primary winding of your induction coil, and from there to the interrupter on the engine, and then back to the battery. From what is called the two secondary binding-posts on your induction coil two wires are run, one to each electrode. The action is as follows: When the engine operates the interrupter a spark will jump from one to the other of the stationary electrodes, and this spark will ignite the charges.

The igniter with movable electrodes has generally one electrode which is mechanically operated by the engine. This style must have, just as the other, an electric battery, but instead of an induction coil, must be provided with a spark coil. The current runs in this case from one end of the battery to one binding-post on the spark coil, then through the coil to the other binding-post and from this binding-post to a switch, and then to a binding-post on the engine and so to the stationary electrode. The

Internal Combustion Engines.

movable electrode is connected to the other end of the battery in such a manner that when a movable arm inside the cylinder touches the fixed electrode the current runs from the battery through the spark coil and the fixed and movable electrodes back to the battery.

With this arrangement an electric spark will occur between the two electrodes when the action of the engine separates them. When we describe the different engines we will go into this more in detail, so that you will be able to locate your trouble and fix it, if it should occur.

One or the other of the above kind of igniters must be used to ignite your charge. After the charge has been ignited the combustion takes place very rapidly and is very complete. It is, in fact, an explosion, and creates high pressure as well as high temperature. To give you an idea how high the temperature rises I can tell you that if this temperature was continued for some time it would melt your cylinder, or, in other words, it is about as high as in a blast furnace. If you measure the heat by a thermometer, this will register about 3000° Fahrenheit. Of course, you understand that it is this high pressure and heat which gives the power.

By considering what I have told you above, it is clear that in order to get any power from the internal combustion engines you must introduce a charge, or mixture of fuel and air, in its cylinder, and after having compressed this charge to the proper amount you must explode or fire it. It follows without saying that you must provide suitable means so that you can control and use the power thus created. After the charge has exploded the cylinder contains only products of combustion, and in order to prepare room for a new charge those products must be expelled. Now the same operations are ready to be repeated. Such a series of operations, always in the same order, one after another, is what we generally call a cycle, and as it is more convenient to use this word than to describe the operations each time, we will employ it hereafter.

The above furnishes us with the reason why a single-acting internal combustion engine cannot get more than one power impulse for every fourth stroke of the piston, or, what is the same, every two revolutions of the fly-wheels.

We will now see what takes place during each stroke of the piston, beginning at the time when the piston is close to the cylinder-head and when the crank and the crank-rod are nearly in a straight line. The piston is supposed to be moving towards the fly-wheels and from the cylinder-head. This is the beginning of the cycle, and we will call this the—

First Stroke. The result of this motion is that a slight vacuum or decrease of pressure inside the cylinder is created. This will open the air valve, generally against a weak spring, admitting air to the cylinder. At the same time a cam or a crank has opened the gas or gasoline valve. The two streams of air and fuel meet at the entrance of the cylinder and a mixture of them in proper proportion enters the cylinder, which is filled with this charge, as the piston has moved to the other end. When this is reached the first stroke is ended and the piston proceeds to start on the—

Second Stroke. As soon as the first return stroke is commenced the engine closes the fuel valve, and the air valve is also forced to close by the increasing pressure of the charge and with the assistance of a weak spring. As the piston proceeds towards the cylinder-head the compression increases, resulting in an increase in the temperature of the charge. These conditions continue to the end of the stroke. At the commencement of the—

Third Stroke, before the piston has travelled any appreciable distance, this charge or mixture of air and fuel under pressure is ignited and an explosion takes place. This drives the piston forward with increasing speed until the end of the stroke, and the power is during this time given off to the fly-wheels, increasing their speed a little. Part of this power stored up in the fly-wheels is returned to the engine during the next three strokes.

When the piston reaches the end of its travel in this direction it begins the second return, or the—

Fourth Stroke. The engine opens at once the exhaust valve. Through this valve the burnt but still hot gases are now expelled into the air by the moving piston.

When the end of this stroke is reached all the burnt gases have been expelled. Now the exhaust valve is closed and the cylinder is ready to take in another charge and thus commence the next cycle.

The above cycle was invented by a man named "Otto," and is called after him the "Otto cycle." I might say that without this invention we would probably not to-day have any practical internal combustion engine. Nearly every one is now using it.

I have given you an idea how the internal combustion engine uses its fuel. I will now tell you something about the different fuels which are used in those engines.

Coal Gas.—This is manufactured, as you know, in nearly every community of any size. The raw material is coal. It is used mostly for both public and private lighting and is sold by the gas companies. A good engine will use from 17 to 20 cubic feet of this gas for each horsepower during one hour. On account of different grades of coal as well as different ways of making the gas it is not safe to say that the same engine will use the same amount in two different places. The above is an average and is close enough to estimate on.

Natural Gas.—This is a natural product and is taken

from the earth through deep wells. It is found in several localities. As we could expect, its composition varies even more than the gas manufactured from coal. In some places engines use only 11 cubic feet per horsepower per hour, but in others the consumption is even higher than if coal gas were used.

Producer Gas, also called water gas, is manufactured from cheap coal in a very simple way. It is almost the same gas as you get when you shut the draft off from your furnace and open the furnace door. It burns with a blue flame and does not give any light. It has been used for light in some cases by mixing with it some other gases of high illuminative power. Careful tests have shown that a gas engine can produce one horsepower per hour from the gas produced by burning one pound of coal.

Blast Furnace Gas.—This is a waste product from the blast furnace and not of much importance for you.

Gasoline.—This is a distillation product of crude petroleum. Of all the fuels, this is the most important one, as it is used a great deal for stationary engines and almost exclusively for the portable or traction engines of this class. It is commonly known as "Store gasoline" or "74° gasoline." A good engine will burn from one-eighth to one-tenth of a gallon per horsepower per hour. Gasoline is very common and almost every store is selling it. Nevertheless it is very explosive, and cannot be handled with too great care. This must not be forgotten by the young engineer, or he is sure to get into trouble some time.

If you are using gasoline keep your supply well and securely closed, and see that your tanks and pipes are free from leaks. The best way is to have every joint soldered and not depend on a common pipe joint. The combination of small leak and a match has often proved a disastrous one.

Another thing to remember is to be sure that you always fill your tank in daylight. Do not try to do this when you cannot see. Even in daylight be sure there is no fire of any kind and that no one who is liable to strike matches is near you. Make this an absolute rule.

I will now tell you what is a good way to fill your tank. Secure about six feet, or more if needed, of one-half inch rubber hose. Roll up your gasoline barrel on something so that the bottom of the barrel is higher than the top of your tank. See that the bunghole is on top, and feed in your hose through the bunghole.

When you are doing this see that there are no kinks in the hose and that the outside end is open so that the air can escape. When you have only a few inches left push the hose in as far as you can reach, if possible under the gasoline surface, and pinch the end of the hose so that neither air nor gasoline can get out. Draw your hose out slowly, keeping tight hold on the end until you have brought the end below the bottom of the barrel. Put the hose in your tank, and if you have done just as I have told you, you will have no more to do than to let the gasoline run. It will continue to run as long as there is any left in the barrel or as long as the end of the rubber hose inside the barrel is under the surface of the gasoline.

Coal Oil or ordinary "lamp oil" is, like gasoline, a distillation product from crude petroleum. Very few engines can use this fuel.

Alcohol is mostly a distillation product of wood, and is hardly ever used in this country as fuel for internal combustion engines.

DESCRIPTION OF STATIONARY GASOLINE ENGINES.

Our first cut (see Fig. 48) represents one of the "Otto Gas Engine Works" machines. We have selected this as a typical stationary engine and will describe it in all details, as completely as possible.

Now look carefully over this cut, so that you will remember it, as we shall have to refer to it a great many times. We note first that very many parts are quite familiar and similar to those of the steam engines. As we have mentioned before, this engine has two very heavy fly-wheels, and we saw also why this is good practice on the internal combustion engines.

Fig. 48.

Internal Combustion Engines.

I want you to note the size of the crank shaft. It is considerably larger than the corresponding part of the same size of a steam engine. If you remember the weight of the wheels and the very heavy pressures the crank is subject to, no more need to be said on this account.

The main bearings are for the same reason also larger.

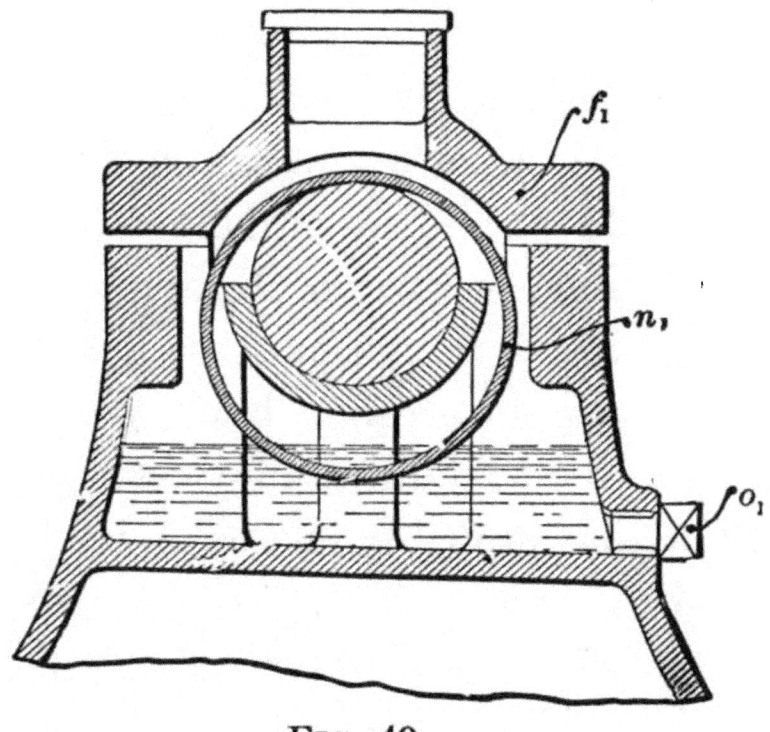

Fig. 49.

The oiling is taken care of by an arrangement (see Fig. 49) consisting of a small ring dipping in an oil reservoir and hanging loose over the main shaft. As the shaft revolves the ring is carried round with it. This supplies all the oil the bearing will need. By unscrewing the plug o you can drain the reservoir. I would advise you to do this once a month. New oil is supplied from the top. All the attention this kind of oiling device needs is to see that the ring is moving, and occasionally to add some new oil. See that the covers are always on, so that no dust can get in.

The piston is the next part of importance. You can see that this is very much longer than the one we would use in a steam engine of that size. It has generally three packing rings in place of the two in the steam engine. This is necessary on account of the very high pressure that the piston has to stand. The rings are made in the same manner as for the steam engine. They are, as you know, of cast-iron and must be handled with care, as they are very easy to break. I will mention here that if you notice any black oil coming out from the cylinder, you will find that probably one or more of your rings are broken, and then you must investigate as soon as possible. Remove your piston and replace the broken ring with a new one. The sooner this is done, the better for the engine. Keep an extra piston ring always on hand. It may save you or your employer a good many days and it costs very little. You may not need it for years, but when you do, it pays not to have to wait for the factory to send you one.

The single-acting engine has no piston rod. The crank **rod** pin is generally supported in the walls of the hollow piston. (See Fig. 50.) It is oiled from the cylinder oiler when the engine is running. Before you start your engine you must turn your fly-wheels so that the small

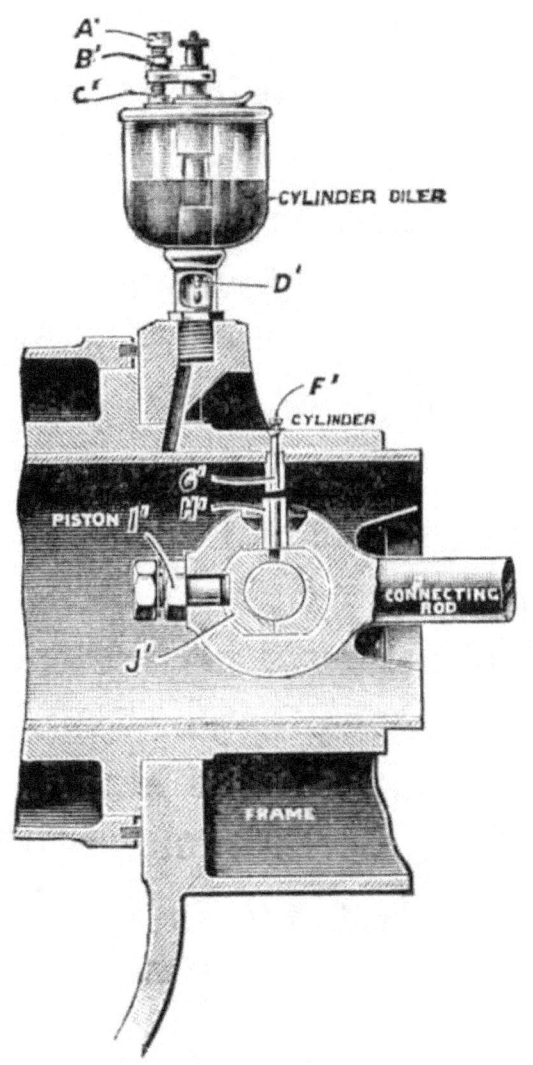

Fig. 50.

hole in the top of the piston and the oil hole in the top of the cylinder are in line, and then supply some oil by squirting down through both holes. (See Fig. 50.)

I will tell you now, don't use any other oil in the cylinder oil cup than high test oil, and never filtered oil. By remembering this you will escape a lot of trouble.

The crank pin is supplied with oil from an oil cup on top of the guard, by means of a wiping arrangement similar to those used on the steam engines. This is another cup I would advise you not to use filtered oil in.

Cylinder.—At first sight this important part looks exactly like the steam cylinder, but by closer inspection you soon discover differences. Note on our figure 48 that the top of the cylinder has an opening; or you may turn to page 201 and look at figure 54 and note the pipes B, A, and D. You will see that A is the inlet, B the outlet, and D the waste pipe for cooling water. From this you conclude that what at first you regarded as the cylinder is in reality only the outside shell, and that between the real cylinder and this shell is a space intended for cooling the cylinder when the engine is running. The cuts, figures 51, 52, and 53, give you a very good idea how this important part of an internal combustion engine is made.

Figure 51 shows the cylinder bolted to the frame of the engine. The long stud bolts hold the cylinder head in place. Asbestos packing about $\frac{1}{32}$ inch thick dipped in linseed oil keeps this joint tight. If this joint should start leaking, you will have to remove the cylinder head, scrape the joint clean from the old packing, and put in a new one. Figure 52 shows the cylinder head with the valves removed. Figure 53 represents the outer shell, between which and the cylinder the cooling water circu-

FIGS. 51, 52, and 53.

lates. Water is mostly used for cooling, but some manufacturers use oil.

Stationary engines are generally supplied with water under pressure from some water system. In this case the water enters at a point marked A (see Fig. 54) close to the exhaust valve, and flows between the shell and the cylinder. It leaves the engine on top through a pipe marked B in an open stream. A funnel-shaped cup, cast in the shell, and provided with a drain-pipe marked D, carries the hot water away.

The valve G on the same figure regulates the amount of water. Enough water must be turned on so that the temperature of the water leaving the engine is between 130° and 190° Fahrenheit, or just hot enough not to burn the hand.

If water is scarce, the same water can be used over and over again by providing a large water tank in which the surface of the water is higher than the top of the cylinder. Near the top and the bottom of this tank pipe connections are made to the cooling jacket of the engine. The flow is automatic and depends on the fact that hot water is lighter than cold. When the water in the cylinder jacket has been heated, it is forced to rise to the top of tank by the heavier cold water. The water in the tank is cooled by the air, and thus the circulation is kept up. By this cooling system the amount of water required is, however, too large for use with portable or traction engines. In order to reduce it, artificial cooling is resorted to. By the use of this system some manufacturers claim to be able to run on a total of eight buckets of water in the tank and a supply of four buckets per day.

Internal Combustion Engines.

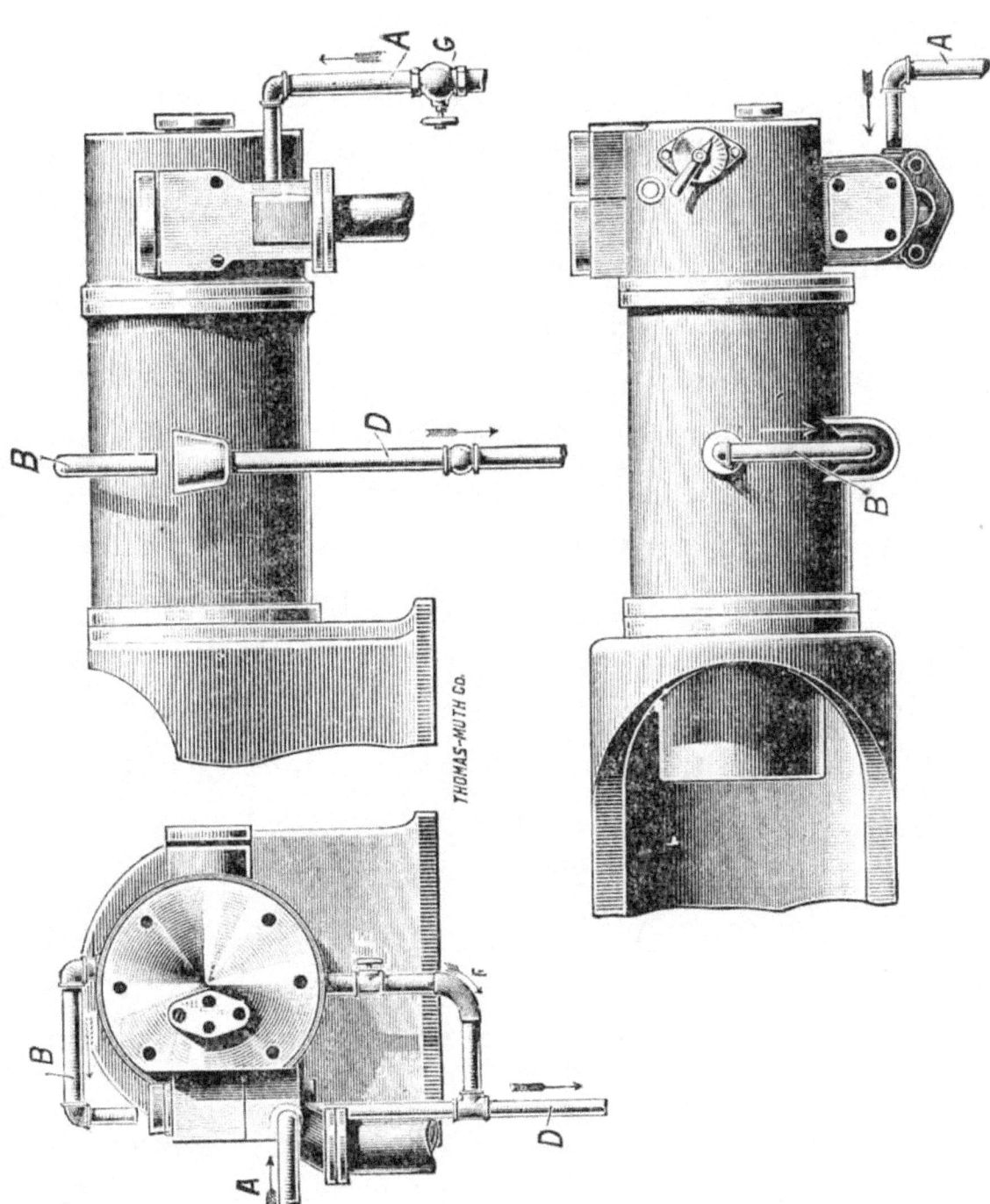

Fig. 54.

The purpose of all these arrangements is to keep the temperature of the cylinder walls considerably lower than the gases. You remember I told you that the temperature of the gases inside the cylinder of an internal combustion engine is about 3000° Fahrenheit, or about as hot as a good clean fire under a boiler. If no means of cooling were provided, the cylinder would therefore become red hot in a very short time. This would soon stop the engine from running, and might even result in a total wreck of the engine. Therefore always keep an eye on your water. Never forget it. If there is the least possibility that the supply is not reliable, provide yourself with one or two barrels and arrange so that you always have them filled with water. Put them up high enough so that the water will run through the engine by gravity. Have a hose attached so that they are ready for use at once in case of an emergency.

Never let your engine run longer than five to six minutes without water. If water is scarce, you can save some by letting the temperature of the overflow get up to 200° Fahrenheit—but watch your engine, as you will be taking considerable risk.

Underneath the cylinder you will find a valve marked E on a pipe connected to the drain-pipe D. This is intended for draining the water between the shell and the cylinder. Always run the water out of your engine when you stop down for any length of time. It is absolutely necessary to remember this in cold weather. If your water-supply is hard or very muddy, daily washing out is necessary in order to keep the space between the shell and the cylinder clean. To allow this space to be filled up with scale or mud would have a very serious result, as that part of the cylinder walls which the water could not reach would soon become overheated, causing premature ignition with a probability of wrecking the engine. Once a year it is good practice to wash this space out with a weak solution of muriatic acid and water.

There is another thing about the cylinder I want to call to your attention. Put a spirit-level on the side shaft and you will see that it is not level; it leans towards the fly-wheels. Now, that is just what it ought to do, and to put up the engine any other way would produce bad results. The reason is that you must prevent as much as possible the lubricating oil from getting into the cylinder, and at the same time supply the piston with enough oil so that it will not cut. The oil cup for the cylinder is generally of the sight-feed type, and for medium-sized engines from two to three drops per minute is generally enough, except when the engine is new. Here is a chance for a careful engineer to use good judgment.

What harm could it possibly do to get lubricating oil in the cylinder? Well, lubricating oil is largely composed of carbon, and when you burn it you know that it will make a great deal of soot. This is just what happens when lubricating oil gets into the cylinder. The soot is deposited on the walls and on the end of the piston. On the walls it does not do a great deal of harm; every stroke of the piston cleans the soot off; but even here it may result in premature ignition. The most danger is due to the soot deposited on the end of the piston. If this becomes thick enough, it will get on fire and ignite the charge before the piston has reached the end of its stroke. Such a premature ignition may result in a cracked crank, a bent crank rod, or a leaky packing. It is the engineer's duty to guard against such possibilities. Only carelessness is responsible for such an accident. This does not mean that you must examine your piston all the time. The engine will let you know, if you pay attention to it, when it needs help. If you have a black smoky exhaust, or if your igniter electrodes need much wiping off, it means that you are probably feeding too much oil to the cylinder, and if you reduce it and get better results you do not need to be afraid of carbonization. Now don't think that this means that you must tinker with the engine all the time. It doesn't. Let well enough alone. Remember this, as it is a good rule for all kinds of machinery.

Once more, we note on figure 48 a long side shaft extending all the way back to the cylinder head. It rests in bearings (see Fig. 52). This shaft gets its motion from the crank shaft by means of a pair of spiral gears. I will caution you here, never to remove any gear wheels from a gas or gasoline engine, for repair, cleaning, or for any other cause, without first marking two teeth, one on each wheel, or, better still, mark one tooth and the groove in which it belongs. Most manufacturers mark them before the engine leaves the shop; but it is necessary to know that your engine is so marked. Therefore do not remove any gears before you are sure. When you put them back, see that you get the marks together. If you don't, you will have trouble. You can readily understand that if the cams and valves which are operated from this shaft don't move at the right time, your engine will not work at all, or will work badly.

Governor.—The first purpose of this side shaft is to drive the governor. To accomplish this, a bevel gear wheel is securely keyed on it. (See Fig. 55.) This wheel engages with a small pinion which drives the governor shaft. The governor balls move out in accordance with the speed, just as in a governor on a steam engine. Instead of operating a valve rod directly, the governor moves a little wheel or gasoline roller, J, from one side to the other by means of an angle lever, I and D, provided with a forked end.

204

The Traction Engine.

Sometimes this gasoline roller runs over the cam K on the driving shaft, and sometimes alongside of it. When the gasoline roller J runs on the cam K, it lifts the lever W, which opens the gas or gasoline valve so that the engine can take in fuel. When it runs on the side of this

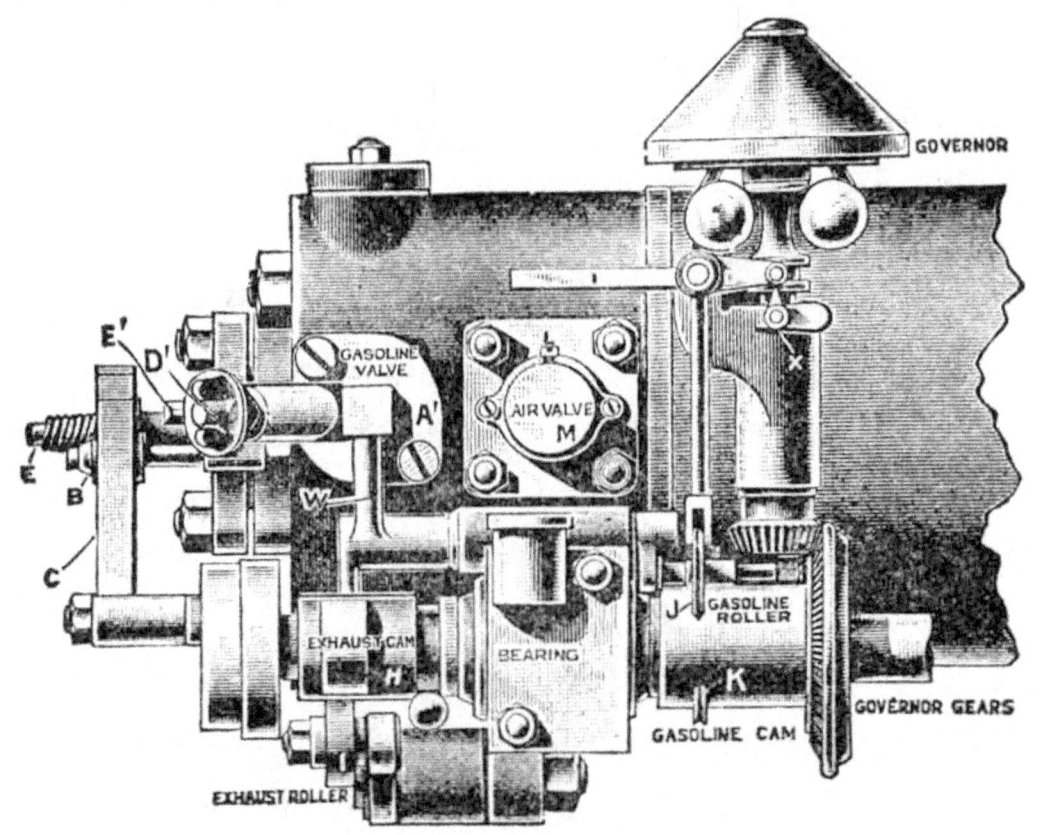

FIG. 55.

cam K, the supply valve is closed. The speed of the engine depends, therefore, on the supply of fuel. When it does not get enough gasoline, because the valve is shut, it will slow down, and this brings the roller J over the cam K which opens the gasoline valve. The engine will then increase a little in speed, causing the roller to be moved back out of the road for the cam K.

This way of governing the engine has been called the "hit and miss way," or the "hit and miss principle." The cover on top of governor can be lifted up in order to oil the levers. By changing the weight of this cover, the engine speed can be slightly changed.

Fuel.—Different kinds of fuels must be supplied in different ways. If gas is used, it is only necessary to pipe it to the engine. The precaution is generally taken to introduce in the pipe line a rubber bag, or some other device which will act as a reservoir. There is no need of any special devices to furnish the gas except to see that the pipe is large enough. If gasoline is used the case is different. For the sake of safety, the gasoline supply is generally located outside the building and on a lower level than the engine. It must therefore be pumped up when needed. For this reason (see Fig. 56) the engine is provided with a small plunger pump P, which is so arranged that it takes the gasoline from tank T through pipe S and forces it up to the overflow cup N. From this cup flows such gasoline as is not used back to the tank T through the overflow pipe O. The pump P is operated, when the engine is running, by the eccentric mounted on the side shaft and secured to the pump plunger I by the wing stud M.

If the engine is left standing any length of time, the

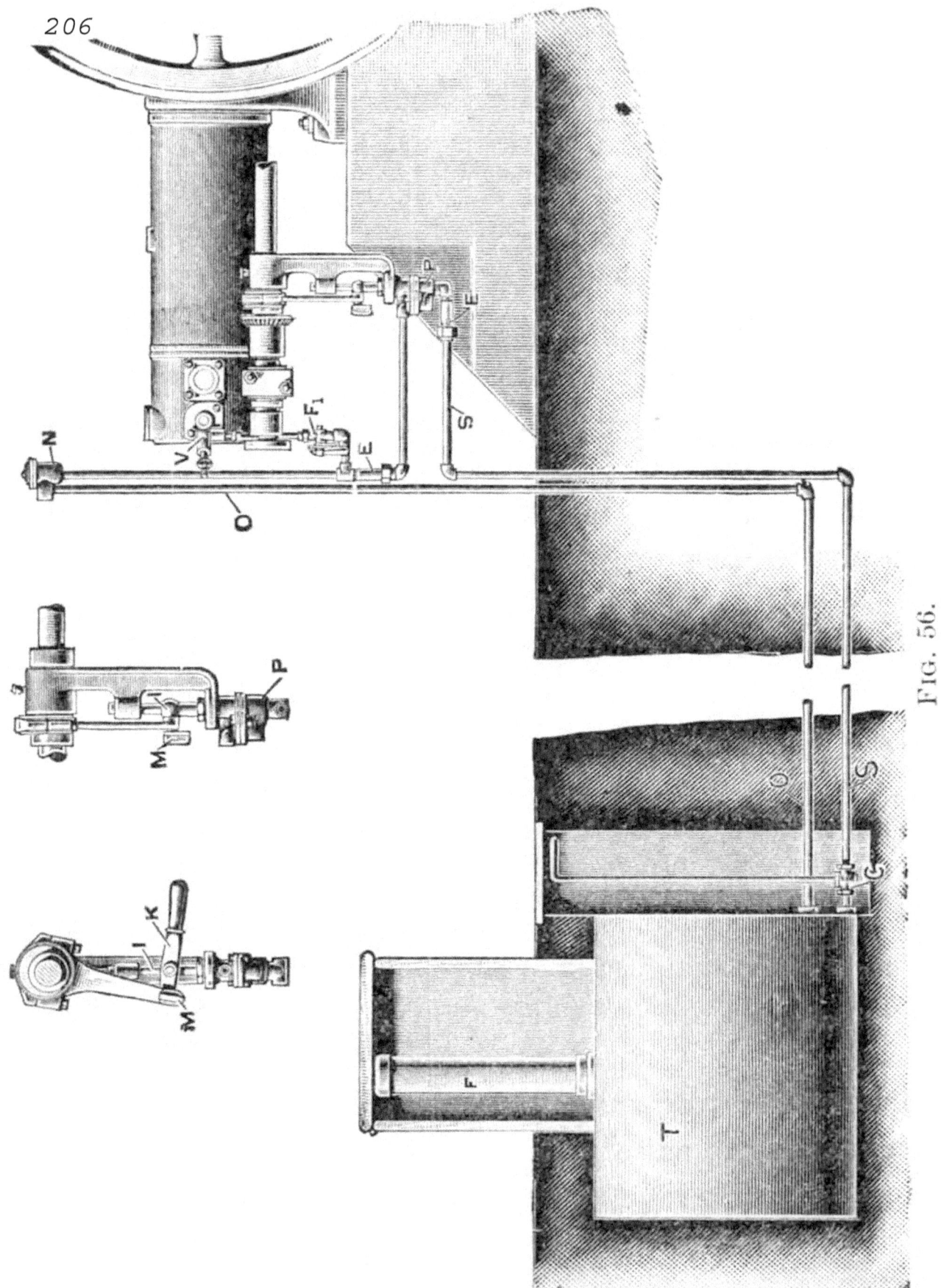

Fig. 56.

gasoline has generally leaked back to the tank T and a new supply in the overflow cup N must be secured before you can start again. Under those conditions you remove the wing stud M from pump plunger I and insert the centre stud on the small handle K supplied for this purpose in the hole of the pump plunger I. Screw in the wing stud M in the eccentric rod, insert the stud M in the hole on the handle K and the pump can then be operated by hand. After you have pumped up enough gasoline in the overflow cup N, so that it overflows in the pipe O, place the pump in condition to be worked by the engine by reversing the operation just described. After you are through with your small hand-lever, put it back in its proper place, so that you do not have to hunt for it next time you need it. Give the plunger a drop of oil and wipe the pump clean. Attend to this before you leave the pump.

The main valve controlling all the supply of gasoline to the engine is located on the supply pipe a little under the cylinder head and is marked F_1 on figure 56. This valve is generally of the old conical type and has a tendency to leak. A little emery and oil can easily make it tight. A leaky valve is not a good recommendation for the engineer, and besides it is dangerous; therefore see that it is tight.

This valve is furnished with a handle, one end of which

plays over a graduated scale, so that the position of the valve can be located. The other end of the handle carries a small electric switch, and plays over a contact piece fastened on an insulated support. This contact piece has a binding-post on one end, and to this one of the battery wires is connected. When the valve is opened the small electric switch on the handle makes contact with the contact piece. When the valve is closed, the contact is broken. After passing this valve, the gasoline flows through a small pipe to another valve marked V on figure 56. This last valve is of the needle type with a fine thread on the regulating stem, so that the amount of gasoline can be regulated very accurately. It is adjusted in the shop and ought not to be changed except by a man who knows his business. After the position is determined a mark ought to be made on handle D (see Fig. 55) opposite the little rod E so that it can be moved back to the right position in case it should have been disturbed. It is a good rule not to move this valve at all. The engine is very sensitive, for too much or too little gasoline and one-half a turn on this valve may bring it out of adjustment. If on your engine there is no mark, make one at once, so that you can see if anybody has meddled with it or not.

Before the gasoline can enter the cylinder it must pass another valve, the gasoline valve proper. This valve is of the poppet type and is operated through the lever W

Internal Combustion Engines.

by the governor. (See Fig. 55.) A spring secured to the exhaust lever stud keeps this valve always shut, except when the small gasoline roller J runs over the cam K. (See Figs. 55 and 57.)

Exhaust Valve.—(See Fig. 57.) This valve is also of

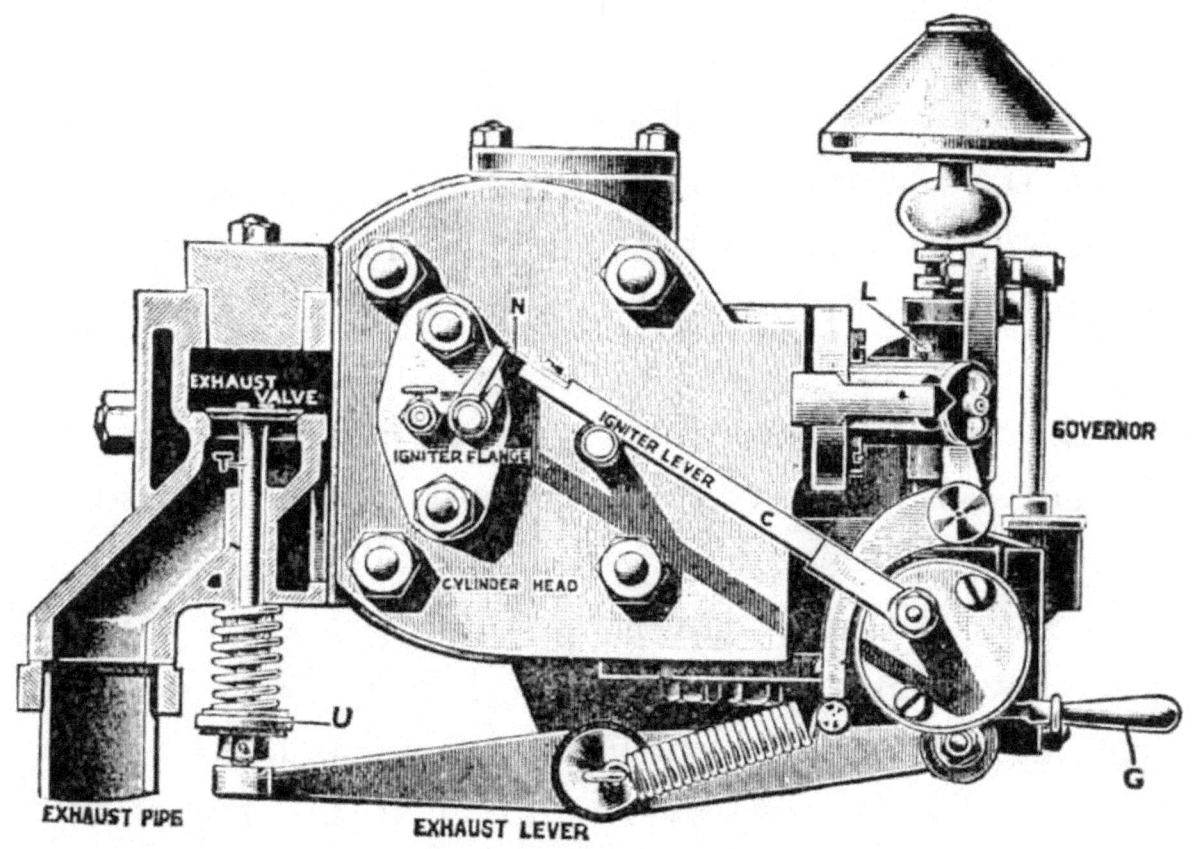

Fig. 57.

the poppet type and should always be tight, a strong spring keeping it shut. It is operated by a cam on the side shaft through the exhaust lever. The construction is plainly seen in the cut. The exhaust lever carries a small steel roller on the end which rests against the cam H. (See Fig. 55.) The same spring which keeps the exhaust valve shut also keeps the roller tight against the cam H. You can see from the cut the action of the cam H on the exhaust lever. Every time the cam H passes over the small roller on the end of the lever the spring under the exhaust valve is compressed and the valve opens. This valve is always hot, and I advise you not to use oil on any part of it. The oil will only make the valve stick harder. If it does not work properly, take it out and clean it with coal oil and put it back again. If it shows any bright spots, rub them with fine emery cloth.

The cam H on the side shaft is put on a loose sleeve which can be moved by a small hand lever along the side shaft. If you look carefully at figure 48 you can see this lever with the handle just under the gasoline valve. As shown in the picture, the handle is in position for starting the engine. On figure 55 it has been removed in order to show the exhaust valve clearly. The sleeve carries two cams, and by pulling the lever to the position as in figure 48 a larger cam comes under the roller of the exhaust lever, this making it easier to start. When the engine begins to run, push this lever back towards the flywheels as far as it will go and leave it there.

Air Valve.—This is placed next to the gasoline valve. (See Fig. 55.) It is like most of the valves of the poppet

Internal Combustion Engines.

type and is kept shut by a small spring. It works automatically, being opened by the suction of the piston and closed by the spring and held closed by the compressed charge.

Igniter.—(See figures 58, 59, and 60.) Figure 58 gives a view of the igniter removed from the machine. Figure 59 is a cross-section, and figure 60 is a top view

Fig. 58.

of the same. It consists mainly of two parts, a stationary electrode marked A and a movable one marked I and O. The stationary part consists of a stem, A, carefully insulated. The insulation consists of two lava bushings, marked C. Small asbestos washers are used on each side of the lava bushings and the whole is tightened up by the nut G, which also serves as a binding-post for one of the wires from the battery. The end inside the cylinder carries a small platinum washer. The electric spark jumps from this washer to another small piece of platinum on the end of the movable electrode.

The movable electrode is not insulated from the engine.

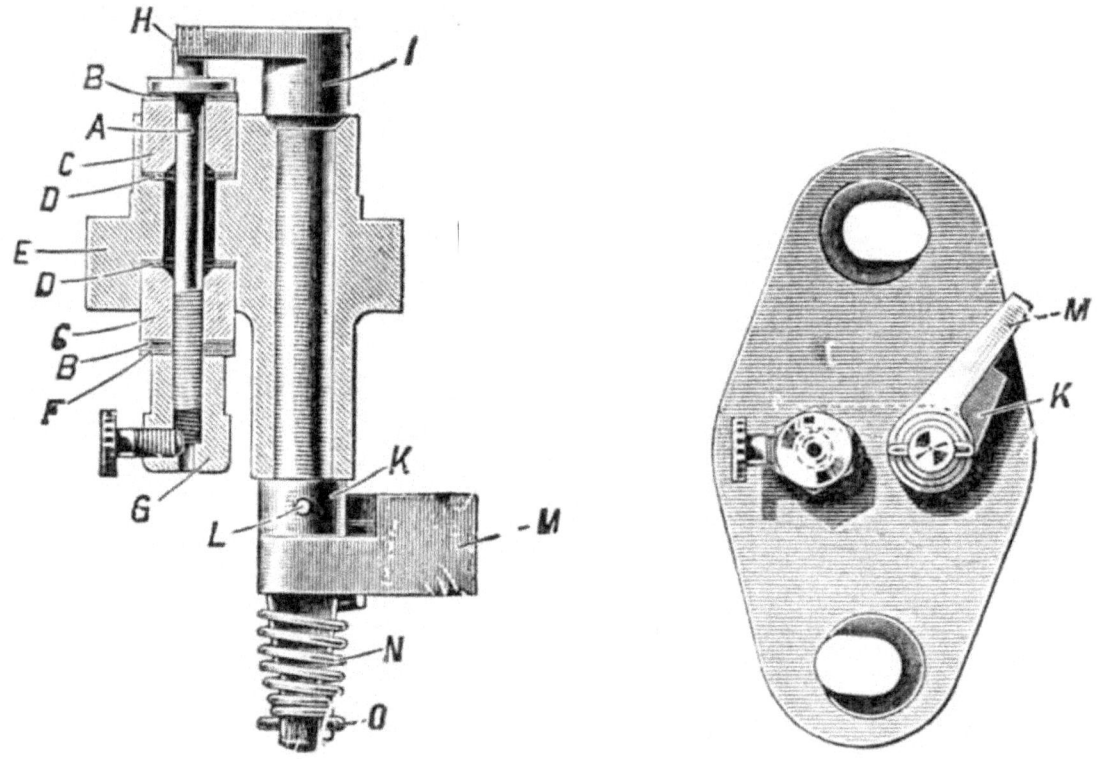

FIGS. 59 and 60.

This electrode consists of a long stem with a valve-like shoulder which, inside the cylinder, has the shape of a small lever, on the end of which a small piece of platinum is fixed. The other end carries a piece marked K pinned to the stem. A movable piece M is connected by the spiral spring N to this piece K. This brings the lever

Internal Combustion Engines.

I and the platinum piece on it against the platinum disc on the stationary electrode without a blow and with an even pressure. The current runs from the insulated contact piece under the gasoline valve handle, through the handle and the iron-work direct to this electrode.

The movable electrode is operated by the igniter lever C (see Fig. 57), which, pushing against the piece M, brings the two platinum pieces together and starts the current flowing. The action of the spring N separates them.

The igniter lever is operated from the side shaft by being hinged on an eccentric stud, secured in a disc on the end of this shaft. The igniter must be thoroughly understood, and it is well to look it over every now and then. If you find a small leak of gas, it is best to remove the igniter at the first opportunity and examine it carefully. If any of the bushings or washers look doubtful, replace them by new ones. Keep always some of these lava bushings and asbestos washers on hand. They are not expensive, so there is no excuse for not having any in stock. Before you replace the igniter, see that the movable electrode moves freely, and wipe both electrodes clean and dry. A little cylinder oil on the inside surface prevents condensation and will not hurt.

As the joint between the cylinder-head and the igniter is ground, I want to caution you to be sure that both those surfaces are clean before you replace the igniter. It is also important that after replacing the igniter you take up, first on one nut and then on the other, a little at each time, alternating between them until the joint is tight, as it is easy to spring the flange of the igniter by drawing up too much on one nut.

PART ELEVENTH.

HOW TO RUN A GAS OR GASOLINE ENGINE.

Well, that was what we started out to tell you. You may think that all that we have told you so far does not have anything to do with that subject, but it is not so. It has everything to do with it, and the better you remember it, the better you will handle your engine.

We will suppose that it is a gasoline engine. A gas engine is for all practical purposes a gasoline engine, with tank, pump, and gasoline left out. Therefore if you know how to start the gasoline engine, you can certainly start the gas engine. We will further suppose that the engine has been erected by an experienced man, and that everything is in its proper place and that nothing is missing. Now, what would you do? From your knowledge of machinery, you would naturally pick up an oil can and start to fill the cups and the bearings of the engine. Right here I will tell you that an internal combustion engine wants good oil, and that for the cylinder must be used only a special high test oil with a very high flash point. That means an oil which does not burn before it is very hot. Never use a vegetable or animal oil for this class of engine. If you do, you are sure to have trouble. Use only mineral oil. Heavy cylinder oil will do. A sure sign of too light oil is that the exhaust is smoky and that smoke comes out of the open end of the cylinder. Light oil will also cause the packing rings on the piston to stick in their grooves and will deposit carbon on the cylinder and on the piston. This last is probably the most serious trouble of them all.

In order to get good oil for your cylinder tell your oil man that the oil you want must have a flash point of about 400 degrees Fahrenheit; that the burning point must not be below 475 degrees Fahrenheit, and that the specific gravity must be about $24\frac{1}{2}$ degrees Baume, and he will know what you want. If he does not have it on hand, he can get it for you.

Another thing about oil. In order to save it you may filter the old oil and then use it over again, but it is best to make it a rule never to use filtered oil on your crank-pin.

Put also a few drops of oil on the cams and on the bearings and also on the plunger of the pump. After you are sure that every oil hole has got a few drops of oil,

How to Run a Gas or Gasoline Engine.

take some waste and wipe off the excess of oil all over the engine

Remember, you must always keep your engine as clean as possible.

The next thing you want to know, is whether your igniter and your battery are in good working order or not. To find out if the igniter is in good condition, you will have to take it out and examine it, as we told you before. For the present we will suppose it is all right.

To test the battery, take out the wire in the binding-post of the stationary electrode, close the little battery switch, which is placed near the engine, and strike a light blow with the bare end of the loose wire on the binding-post where the other wire is connected. If you get a bright spark the battery is in working order.

We will take for granted that you have examined the gasoline tank and that you have found it full with gasoline and also that all the connections between the engine and the tank are tight.

It remains, then, to supply your engine with gasoline before you can start it. To do this it is only necessary to disconnect your gasoline pump from the engine and operate it with hand. I told you how to do this in the description of the engine, so it is not necessary to repeat it here. You must keep on pumping until the gasoline appears in the overflow cup. You can see if you have enough by lifting the small cover. Sometimes, if the engine has stood long, the pump will not lift the gasoline on account of the air in the pipes. A little gasoline poured into the overflow cup generally starts the pump pumping. You must remember, however, that before you pour in the gasoline you must open the gasoline valve F, see figure 56, a little in order to allow the air in the pipes to escape. Without this no gasoline will reach the pump. Before you start pumping, close it again. As soon as the gasoline has reached the overflow cup, you can stop pumping and must disconnect the pump handle. Connect the plunger on the pump with the eccentric rod so that the pump will be operated by the engine after it is going.

Prop up the governor with the little lever you will find below, pivoted on the bracket supporting the governor. Pull the handle operating the sleeve of the exhaust cams towards the cylinder end of the engine. Put the igniter lever on the smallest diameter of the little roller which supports it. Turn on the oil cups. Pull the fly-wheels round a few turns, so as to clean out the cylinder and to give the oil a chance to distribute itself. At the same time keep an eye on the valves and see that they work properly.

I told you before never to use oil on the gasoline, air, or exhaust valves, but here is a good place to call attention to it again. To keep these valves in good order, I want you to give the stems a few drops of coal oil occasionally. For this reason, keep a small oil can always at hand and see that it is filled with coal oil.

Now note if the small gasoline roller is just in front of the cam which operates it. If it is not, move the fly-wheels enough to bring it in this position. I would advise you to make it a rule always to place your engine in this position when you stop, as you are then ready to start at a moment's notice.

You are now ready to start. Turn on the small electric switch, give the gasoline valve F about a quarter turn, and take hold of the rim of one of the fly-wheels and turn them over as rapidly as you can. Two turns are generally enough and an explosion takes place. Your engine is going. Now open the gasoline valve a little more, push the handle which operates the sleeve on which the exhaust cams are placed, to the front of the engine as far as it will go. Move the little roller supporting the igniter lever in such a position that this lever rests on the largest diameter of the small roller. Turn on the gasoline valve in full. Start the cooling water flowing, by turning on the water valve G, see figure 54, and close the valve E in the drain pipe under the cylinder. When the water is flowing out through the pipe B regulate the stream by the valve G, so that you get about the temperature you want.

Now your engine is ready for a day's work.

Had the engine been burning gas instead of gasoline, you would have started it in the same way, except that all operations which have anything to do with the gasoline would have been omitted, as the gas is supplied under pressure and does not have to be pumped up.

Now suppose that, after you had finished all those preparations, your engine would not start. What would you do? Pick up a wrench and start taking it apart? No, don't do that.

First of all, turn off your gasoline, open your battery switch, attend to the oil cups and stop off the cooling water. Go over the instructions again and see if you possibly have forgotten any. I have seen an engineer work for a long time with his engine trying to get it started, simply because he forgot to prop up the governor.

Why is it necessary to prop up the governor? Well, if you had studied the previous pages carefully, you would not need to ask this question. However, the answer is that no engine can run without fuel, and propping up the governor brings the gasoline roller over the gasoline cam, with the result that the gasoline valve opens when you turn the engine over and allows the gasoline to enter the cylinder. You realize, of course, that the engine will not run if it does not get any fuel.

This brings up another question; namely, does the speed of turning have any influence on starting? When you turn the fly-wheels over, it is clear that you cannot turn them as fast as when the engine is running, and this results in keeping the gasoline valve open very much longer than when the engine is operating normally. This explains why it is necessary not to open the gasoline valve wide, as I cautioned you before, when you want to start up. You can readily see that the longer the automatic gasoline valve is kept open, the more gasoline will get into the cylinder. You might think that a little too much gasoline cannot hurt. Take my word for it and don't try it except if you are very anxious to get some exercise. Your engine will not start with an excess of gasoline, or in case of a gas engine with too much gas; and the longer you keep on trying, the more gasoline you will have to deal with, and your chances for a start are getting smaller and smaller. The best thing to do in a case of too much gas or gasoline is to close your gas or gasoline valve and keep your fly-wheels going for five to six turns, or long enough to be sure that all gas or gasoline has been expelled from the engine cylinder. When this is accomplished, start all over again, but see that you don't forget anything.

In order to be sure that nothing shall be forgotten, it is best to make it a rule to start your engine always in the same manner. That is, whenever you start your engine, do each operation one after another, but always in its turn.

In this way you will get so accustomed to do it right, that you cannot do it any other way.

Now suppose your engine is running all right and that all on a sudden it slows down and then stops. What is to be done? Well, first of all, do what I told you to do in the above case—that is, turn off the gasoline, the oil, and the water. After this is done, examine your fuel supply. In case of a gas engine the rubber bag will show at a glance if you have gas or not. In case of a gasoline engine, you can find out by lifting the cover of the overflow cup. In either case, if you find that you have no fuel, the cause of the stoppage is ascertained. The next step would then be to locate the cause of the trouble. In case of the gas engine, look over your pipes. Probably you will find water in them. Look for a dip in the line, and if the pipes have been properly installed, you will find a drain cock in that place. By opening it and letting the water run out, the cause is removed and you can start again. If you find no water here or in any place between the engine and the meter, notify the gas company at once, as the trouble is outside.

In case of a gasoline engine, you will have to find out first if there is any gasoline in the tank. We suppose that there is. Next you must examine the pump. See that you have no loose connection, or, in other words, that the eccentric is tight on the shaft so that the plunger moves when the engine moves. If this is all right, the trouble must be inside the pump somewhere. Disconnect the pump plunger from the eccentric rod and connect up the pump for pumping by hand. If after a few strokes you do not get any gasoline, you are sure that the fault is in the pump. In order to be able to examine it, you will now have to take it apart. To do this, disconnect the union between the gasoline valve F, figure 56, and the pump, and also between the pump and the tank. You can now swing the pump out round the side shaft so that you can test it. You will find that in the unions which you have just disconnected are located two small conical sieves, one between the pump and the tank and the other between the pump and the engine. If on examination of the sieves you find many of the small holes in them closed up, take them out and wash them clean in gasoline. Now put them back, and closing the pipe to the tank with one hand working the pump handle, with the other see if you can feel any suction. If you do, that part is all right. Try the same experience on the pressure side of the pump, and if you can feel pressure on the hand closing the pipe you can consider that the pump is all right again and that you have removed the trouble. On the other hand, if any of these experiments do not succeed, you can be reasonably sure that the trouble is with the valves in the pump. A small piece of packing or dirt or something similar has probably gotten under the valves. In order to get at them, unscrew the top of the pump, and if everything looks all right wash the whole thing carefully in clean gasoline. If any part should show very much wear, you will have to get a new piece from the manufacturer, and in the meantime you can probably fix it by putting a little fine emery and oil on the valve seat and grinding it in. After this wash carefully with gasoline and try the pump again. Remember that you must not have any fire, matches, lamps, or torches around when you are attending to this. It is best to do this in daylight. After you have gotten everything in place again, try the pump, and if you find that you can feel suction and pressure against your hand, connect up the pump and it will pump all right.

Now suppose that you found gasoline in the overflow cup when you examined it the first time. In that case there was nothing the matter with the pump. And then you must investigate the automatic gasoline valve. Push it in with your hand and see how it works. It it flies back with a snap, it is in good order. If it sticks take it out and treat it as the other valve stems, with some emery on the bright spot and coal oil for lubricator. Before you put it back examine the seat. It is possible that it may need grinding in. If so, emery and oil will make it all right. Wash all emery off carefully with gasoline

How to Run a Gas or Gasoline Engine.

and put it back. If there was nothing the matter with it, try the air valve. Take off the cover and notice if the lock nut on the stem is tight. See that it moves freely when you push it in and that the spring works properly. If the stem needs rubbing down, attend to it in the same manner as with the other valves. When you are sure that it is all right, attend to the exhaust valve. Try it the same way as you tried the gasoline valve. The spring on this valve is very much stronger, so you will have to use more force. Don't leave it before you are sure that springs, stem, and seat are right. Never use anything else than coal oil on all these valves. When you are certain that all the valves are right, the only thing which remains to be examined is the electrical part; that is, the battery, the connections, and the igniter. We have gone over this before, but will repeat it again briefly. The shortest way to find out if this is in good order, is to remove the igniter and to turn on the battery switch and the gasoline valve. The latter only so far that the contact piece on the handle makes contact with the insulated piece underneath. Now rest the igniter tightly on a bright part of the engine and snap the movable electrode a few times. If you get a good spark, it is all right. If not, you will have to examine each part in detail, as we have told you before. After in this manner having located the trouble in one of the following three parts, namely, the battery, wires and connections, and igniter, you proceed, of course, to examine in detail the part which is faulty. For our purpose we will examine them all, one after the other.

Battery.—In order to find out which cell or cells are at fault, we test in the following manner: Take two wires; connect one in each binding-post of the spark coil. Connect one binding post, say the zinc on each cell of the battery, one at a time with one wire and touch the other binding-post of the same cell with the other wire. You will get a spark if the cell is in good order. Try thus each cell, one after another; and if they are good, you will see the same kind of a spark as you saw when you tried the first one. If no spark is visible or only a very much weaker one, we conclude that this cell needs attention. There are three things which may need attention, and they are the zinc, the carbon or copper oxide, and the solution. If the zinc and carbon are very much reduced in size or the copper oxide shows *red* where you make a deep cut with a knife, it is probable that they need to be replaced by new elements. If a sal ammoniac solution is used, very often the water is found evaporated, and refilling the cells with water is all that is wanted.

I might tell you here, that if you note that white salts are appearing on the outside of such cells they need attention. As a rule, a battery ought to last four to five months, according to use, and if no local trouble occurs, all the cells ought to be used up at the same time, so that if it is about this time that trouble occurs, it is best to recharge the whole battery, as in that way you will prevent a great many stops. I want also to caution you not to run bells or any other electrical devices from your engine battery. As a rule, all batteries supplied with an engine are very much more expensive than the ordinary ones. Get a separate battery for your bells or experiments.

Wires, Connections, and Spark Coil.—The most common place where interruption of the current occurs in the wires is at the battery binding-posts. These must be examined carefully; see that the end of the wire is made bright with sand-paper and that the surface underneath the screw-head is cleaned the same way. Be sure that the screws are down tight and that the wire feels solid when you take hold of it. If any joints are made in the wire, see that they are soldered. An ordinary twisted joint is not reliable enough.

Examine your switch and see that all connections are tight and the surfaces on which the switch lever rests are clean and bright. Another point which needs attention is the small contact piece on the handle of the gasoline valve. See that it has spring enough and that the surface is clean. Do not put any oil on the surface; use sand-paper for cleaning. The same holds good for the binding-posts on the spark coil.

What is a spark coil? The spark coil is simply a coil of insulated wire wound in several layers on a bundle of soft iron wires. The action of this coil is to intensify the spark when the current is broken. It is not a very sensitive apparatus, but I will caution you to see that it is not put up in a damp place or where water can reach it. If it is put up in such a place, change it to a dry one, as a spark coil may be destroyed in a very short time if put in a damp place.

Igniter.—After running for a long time this piece may cause trouble by not making contact between the platinum disc on the stationary electrode and the small platinum bar on the movable one. In that case loosen up the outside nut on the stationary electrode and turn the electrode round so that the bar hits the disc in another place. You can also change the striking point on the bar, by adding one or more small thin asbestos washers under the stationary electrode. Before you replace it, see that the nut is tight, but do not screw it up so tight that you break the small lava insulators which insulate the stem from the metal part. Sometimes you will find a great deal of soot on the igniter when you remove it. That is generally a sign of bad oil or too much oil, and needs watching. Wipe it off carefully and rub a little cylinder oil over the surface before replacing the igniter. See if the cylinder oil cup is feeding too fast; if so, reduce the number of drops little by little. The soot may also be caused by too much gasoline. Turn off the needle valve a tenth to an eighth of a turn and watch the result. As a rule, this valve ought not to be moved; but if it is, never move it more than one-eighth of a turn at a time.

I have cautioned you several times against the use of poor or thin oil for your cylinder, and also explained why it is necessary to have good oil, so it is no use saying any more about that.

I have also explained what happens when a packing ring breaks, and warned you not to run any length of time before you replace it. I am now going to tell you what happened to an engineer in New England, who replaced a broken packing ring and then tried to run his engine. After putting the new ring in place, he put the piston back in the cylinder and connected up his crank-rod. The engine started in the regular way and everything seemed all right. After a few minutes, however, loud explosions occurred and the engine slowed up and nearly stopped, but recovered and started again. After a little while the same thing occurred again, with the same results. He realized that something was wrong and stopped. He examined everything very carefully without discovering anything wrong; even took out the piston.

The result was the same. After repeating this experience a few times, an expert from the manufacturer was sent for. The expert examined the engine on his arrival and started it running. His experience was the same as the engineer's, and noting the behavior of the engine he concluded that the trouble was due to premature ignition. On removing and inspecting the piston as well as the igniter, not enough soot was discovered on them to explain the premature ignition. An examination inside the cylinder failed also to reveal any deposited carbon. By a closer examination farther back in the cylinder with the aid of a candle, a small corner piece of a fire-brick was discovered. The piece was found in the extreme rear end of the cylinder head. This little piece of brick was removed. The piston and igniter were replaced and the engine was started. It was now found to run just as well as it had always done.

An investigation of the cause brought out the fact that at the time the piston was removed for putting on the new packing ring, a bricklayer was at work repairing the ceiling over the engine. The small piece of brick had evidently been dropped by him. The engineer did not examine the cylinder closely enough to discover it when putting the piston in place. He evidently pushed the little piece of brick into the cylinder with the piston. After a few explosions this brick piece got hot enough to explode the charge. The carelessness of the engineer in not discovering this piece of brick when replacing the piston was the cause of the trouble. A careful man would not only have inspected the inside of the cylinder, in which case he would have discovered the brick piece, but he would also have wiped out the cylinder with clean waste and afterwards oiled the surface carefully before he would have attempted to replace the piston. During any of these operations he should have discovered the piece of brick, but as he did not, it is very doubtful if he performed any of them. The engineer had been running this engine for many years and was considered a very good man, but it is evident he was not good enough.

If the temperature where you are trying to start your engine is very low, you may have some additional difficulties. If the temperature is low enough, the cylinder may be so cold that the heat due to compression of the charge is not enough for ignition. Under these conditions it is best to get some buckets of hot water and fill the cylinder jacket with it. You will then have no more trouble in starting. Be careful not to turn on too much of the cooling water at a time, and when the engine is running regulate it so that the temperature is about 130 degrees Fahrenheit. To let it get much cooler than that is not advisable, as the gasoline will not gasify well below this point.

After the run is over and when you have stopped down

and fixed your engine so that it will be ready for starting next time, turn off your water-supply and open the drain valve under the cylinder, so that there is no water left standing in the jacket, as this will have very serious results if it freezes. This omission has resulted in broken cylinders and cracked jackets.

PART TWELFTH.

DESCRIPTION OF GASOLINE TRACTION ENGINES.

The gasoline traction engines are generally constructed on the same lines as the steam traction engines; that is, a gasoline engine is placed on some kind of a frame and is connected by mechanical means with the wheels supporting this frame in such a manner that when the engine is running they can be made to revolve or not as wanted. The absence of the large boiler which is a prominent feature in the steam traction engines is the most notable difference. They have the same large and heavy driving-wheels in the rear and the small wheels in front as we are accustomed to see on the steam traction engines. The same steering mechanism is also used. The various manufacturers have in fact made use of the same apparatus which long experience with the steam traction engines has proved to be practicable. In most cases they have adopted their regular type of stationary engines for this purpese, only making such changes in the various parts of the engine as the new conditions required. The "rig" consists generally of a pair of heavy I-beams as foundations, to which the gasoline engine as well as the driving machinery and the axles are bolted. It is generally provided with a rear platform on which the operator stands and from which he can perform all necessary operations. All the various levers and handles controlling the engine as well as the steering-wheel handle and brake lever can easily be reached from this place. If you can operate a steam traction engine, you will have no trouble in handling this type, provided, of course, that you are familiar with the gasoline engine.

THE "OTTO" GASOLINE TRACTION ENGINE.

This engine consists of one of the "Otto" standard single-acting gasoline engines supported on two heavy I-beams. In regard to the engine (see Fig. 61) it is not necessary to say anything here, as it is in every respect like the stationary engine of the same make which we have described in the previous pages.

The engine shaft is connected to the driving-wheels by means of a pair of spur gears. Upon the shaft carrying the two spur gear pinions is placed a straight-faced pulley which is driven by means of friction from a fiber driving pulley placed on the engine shaft. The reversal of the

Fig. 61.—The "Otto" Traction Gasoline Engine.

direction of the traction engine is accomplished by bringing into frictional contact with the fiber driving-wheel on the engine shaft and the straight-faced pulley on the transversal shaft carrying the gear pinions, a small pulley pivoted on a swinging lever below the driving pulley. The driving pulley is provided with a friction clutch which is operated by a hand wheel outside the hub of the pulley.

The arrangement for cooling the cylinder without having to carry an excessive amount of water consists of a small tank (see Fig. 62) holding about ten buckets of water and located above the engine cylinder. It is connected with the water jacket of the cylinder by piping. The water flows from this tank through the pipe connected with the under side of the cylinder into the water jacket, where it is being heated by the hot walls of the cylinder. The heated water, being of a lower specific gravity than the water in the tank, rises automatically into the tank. The cooling of the water in the tank is provided for by forcing air through the water by means of a small centrifugal fan connected by a pipe to the bottom of the tank and driven by a belt from the engine fly-wheel. The manufacturer claims that about two buckets of water are evaporated in this apparatus during one-half of a day's run, or, in other words, that the engine needs to be supplied with four buckets of water per day for cooling purposes.

Description of Gasoline Traction Engines.

Fig. 62.—Water-cooling Device of the "Otto" Gasoline Traction Engine.

No special brake is provided, as all the braking effect needed is supplied by the proper handling of the friction clutch and the reverse clutch lever.

As I told you before, all rules and precautions given you in the previous pages in regard to starting, running, and stopping of gasoline engines are applicable to this engine, and it is therefore not necessary to repeat them here.

HART-PARR GASOLINE TRACTION ENGINE.

This engine, manufactured by the Hart-Parr Company, differs in several points from the previous one, as you can see from figure 63. It is mounted on heavy I-beams and has the same general features which are found in nearly all traction engines. This company furnishes their traction machines with two single-acting engines having their cranks placed 180 degrees apart. They are using the "Otto" four-stroke cycle and are provided with electric igniters. The traction engine is not provided with any special brake. The reversing lever is used for this purpose, as by proper handling of it either friction clutch can be applied more or less. The differential or compensating gear is keyed on the revolving axle. The engines have their cranks and crank rods completely enclosed in order to be able to use the spray type of lubrication. Most of the valves used on those engines are of the poppet

Description of Gasoline Traction Engines.

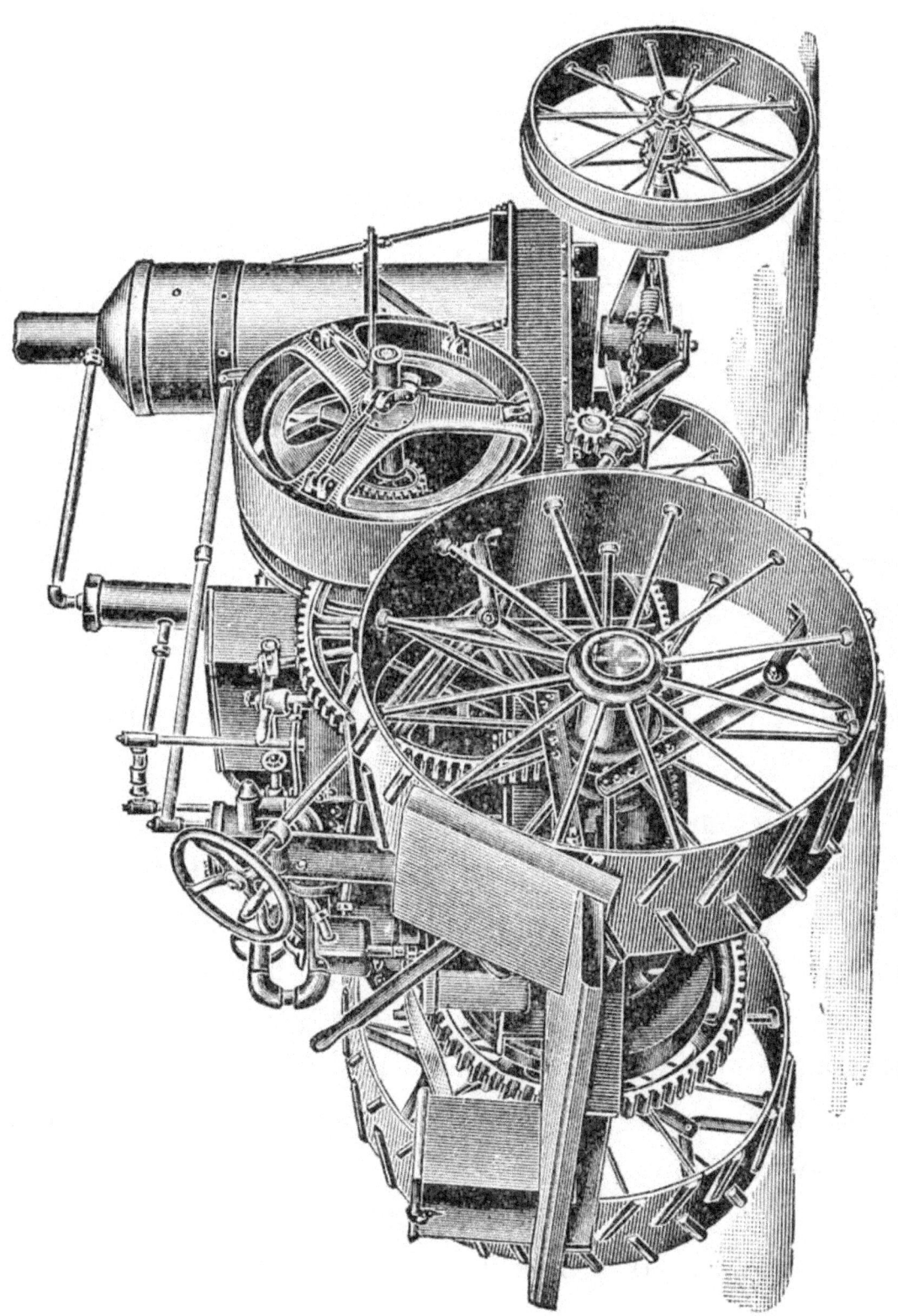

Fig. 63.—Hart-Parr Gasoline Traction Engine.

type. The cylinder-head is of a half spherical form, with removable valves, and has double walls between which the cooling fluid circulates.

For cooling the cylinders of those engines, oil is used instead of water. This apparatus consists of (see Fig. 64) a centrifugal oil pump located on top of the cylinder, which pumps the hot oil from the cylinder jacket into a special cylindrical radiator consisting of a tank having a number of vertical air tubes running through from the bottom to the top. Through those tubes circulates the air, passing in at the bottom and out at the top, thus cooling the oil surrounding the tubes inside the tank. In order to increase the amount of air passing through the tubes, a funnel has been placed on top of the tank, and the exhaust gases from the engines are led into the bottom of this funnel through a pipe like the exhaust nozzle in the steam traction engine and having a similar effect. In this manner the quantity of air flowing through the cooling-tubes is greatly increased and the oil in this tank, which is continually used over and over again, is kept at a temperature very little higher than the temperature of the surrounding air. The above-described funnel acts also as a muffler for the exhaust gases, decreasing the noise of the explosions.

The gasoline tank is located underneath the I-beams and the gasoline is pumped from this tank to the engine by a small pump.

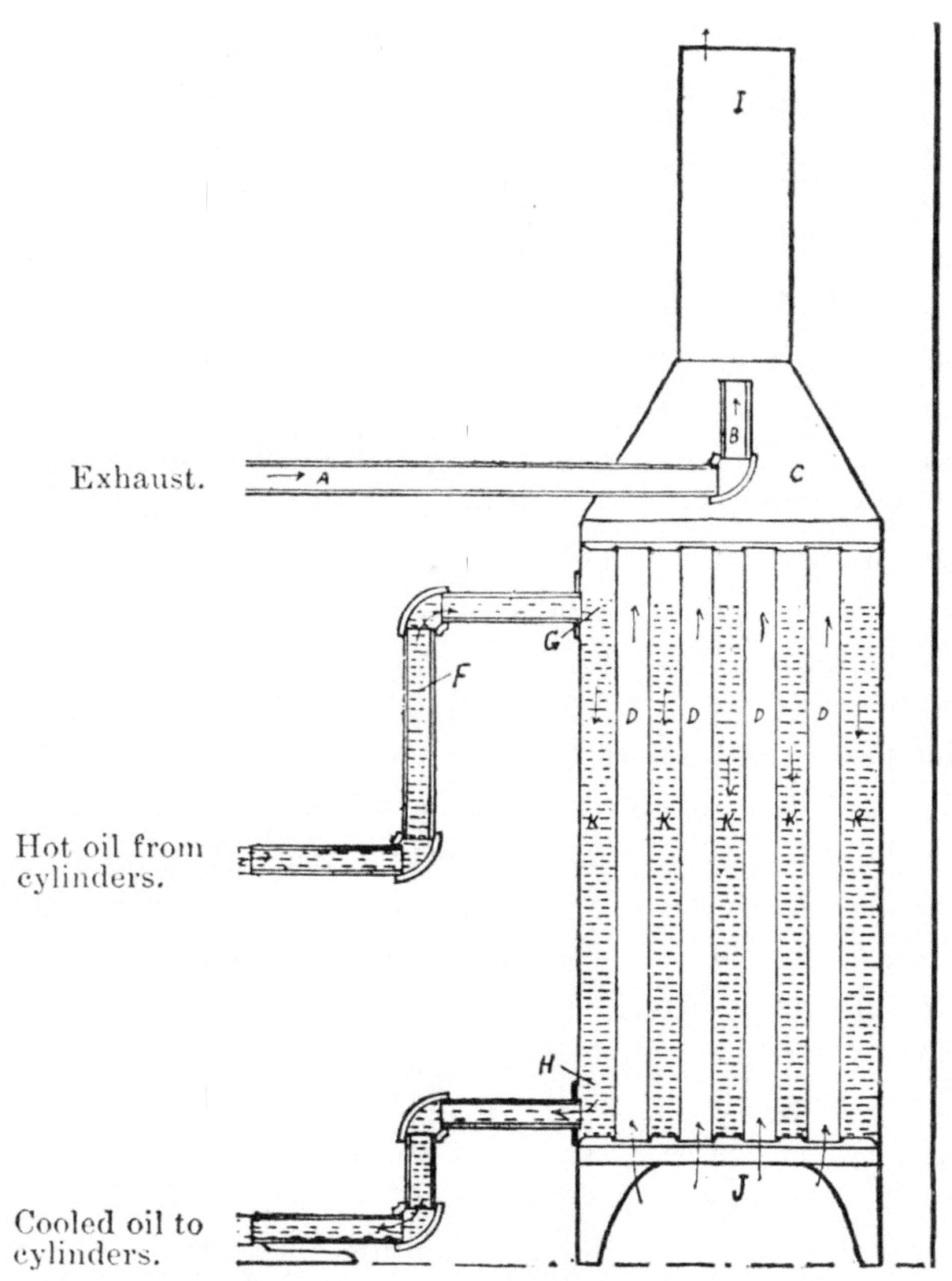

Fig. 64.—Radiator for Cooling Oil in the Hart-Parr Gasoline Traction Engine.

The Traction Engine.

The electric igniter is of the "movable electrode" type. Both a battery and a small dynamo are furnished, together with a switch so arranged that only one or the other can be used at one time. When the engine is to be started, the switch is put over on the battery side and kept there long enough for the engine to get up to speed. When this is reached, the operator turns the switch over to the dynamo side, allowing the dynamo to furnish the current of electricity needed for the igniter. The governor is of the "hit and miss" type and can be changed very quickly to vary the speed of the engine while it is in motion. The power is transmitted directly from the crank shaft by means of a train of spur gears including two friction clutches provided with large wooden friction blocks. In the train of gears on one of those friction clutches is introduced a small pinion, so arranged that when this friction clutch is thrown in, the engine will drive the traction engine in an opposite direction than the one the other friction clutch drives it when it is thrown in. These friction clutches are operated by means of a lever so designed that when placed in one extreme position, the engine runs one way, and in the other extreme position the engine runs the opposite way. When this lever is placed in the middle position, neither one nor the other of the friction clutches is engaged. By slowly moving the lever from the middle position to one extreme, the

Description of Gasoline Traction Engines.

machine will start very slowly and gently; and by throwing the lever quickly over to the other extreme, it will quickly stop and then reverse its direction. This lever acts also as brake-lever if it is applied gradually, and therefore no special brake is supplied with this traction engine.

The driving-wheels, steering apparatus, etc., are similar to the ones used on the steam traction engines, and need no special comment.

PART THIRTEENTH.
THE THRESHING MACHINE.

We suppose that you are familiar with the threshing machine in general. It is a relatively large machine and looks somewhat complicated, though in reality it is quite simple.

We can divide its action into three operations, namely, the threshing proper, the separation of the grain from the straw and the chaff, and the stacking which disposes of the straw.

The first operation is performed by a revolving cylinder, having teeth set in rows projecting from the surface and rotating with a high speed.

Stationary curved parts, called concaves, also having rows of teeth, fitting in between the teeth of the cylinder, partly surround it. The teeth on the cylinder do the threshing, while the teeth in the stationary parts, or the concaves, serve to hold the straw while the grain is threshed out.

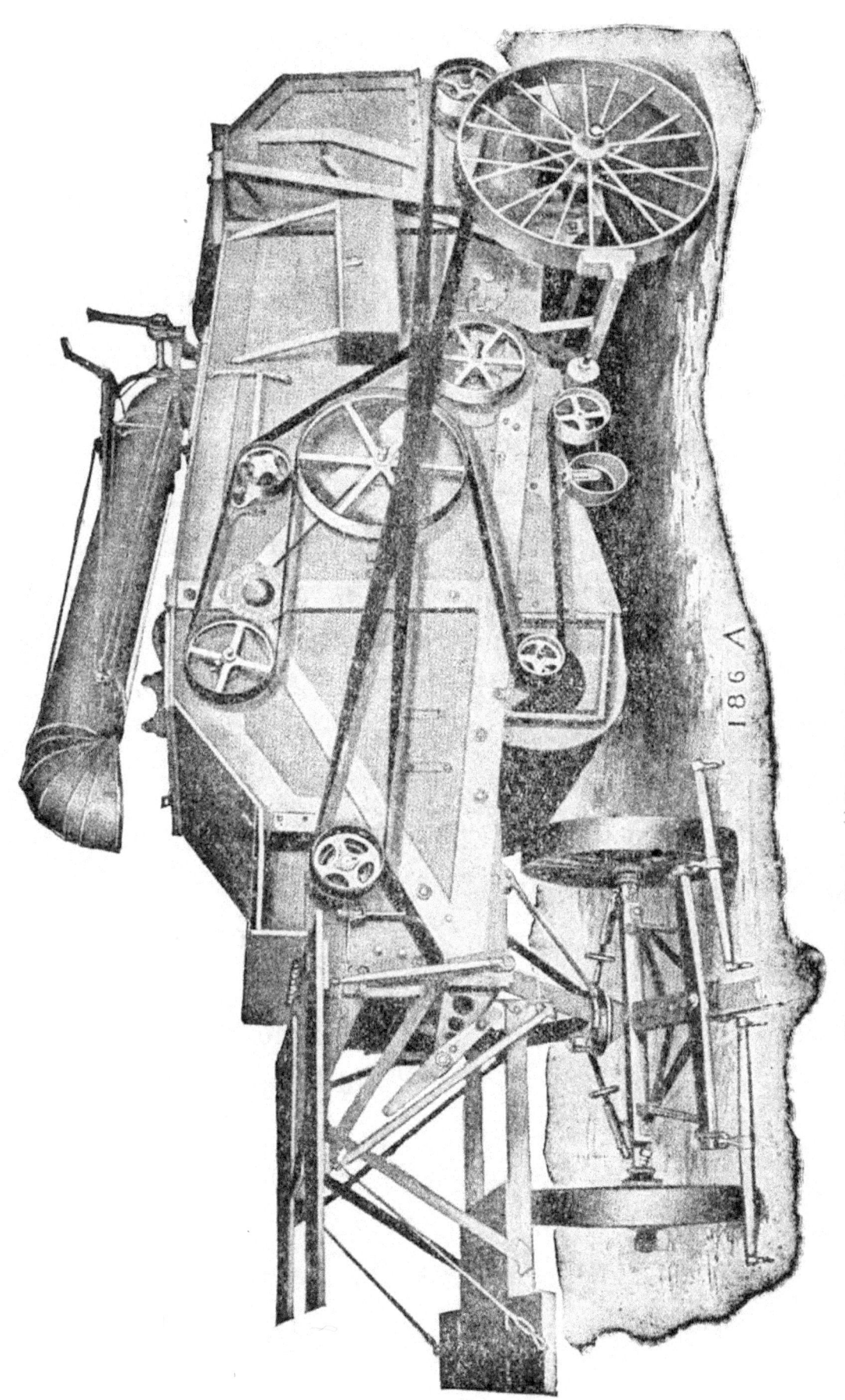

Fig. 65.—The "Landis Eclipse" Thresher.

The second operation, or the separation of the grain from the straw and the chaff, is performed in various ways by the different manufacturers. On this account we can divide the threshing machines into three classes, which we may call the apron type, the vibrating type, and the agitating type. There is very little difference between them, and some machines are even combinations of them all.

The last operation, the disposing of the straw, is performed by the stacker. This part of the machine carries the straw from the thresher and delivers it on the stack, either by means of an endless chain of slats working in a box, or by a blower and a pipe, which combination is generally known as a "wind stacker."

Where the operation of removing the straw is performed by means of a wind stacker, the operator has a better control over the straw than when a common stacker is used.

In addition to the threshing machine, and sometimes combined with it, are often found baggers, with or without tallying attachment, weighers, with or without conveyor; wagon-loaders, grain registers, dust collectors, pneumatic grain elevators, band cutters, and self-feeders. All of these machines are easy to handle and need no special descriptions.

The Threshing Machine.

GENERAL DESCRIPTION.

Figure 66 represents a longitudinal section of a "Landis Eclipse" thresher and shows the relative position of the parts.

The straw enters from the front of the machine, over the grate, and passes between the cylinder and the concaves. It is here subjected to the beating action of the rapidly revolving teeth on the cylinder, which loosens the grain from the husk. The free grain falls direct from here through the concaves or the separating grate to the "grain bottom." The grain which still remains in the husk is separated by being thrown with great velocity against the deflecting plate located close above the cylinder. As the grain is liberated, part of it falls directly to the grain bottom below and part of it falls on the straw and is carried away with it. The straw, helped along by the action of the picker and the straw racks, is moved forward under continual vibrations and delivered to the stacker in the rear of the machine. The continual vibrations of the straw racks shake out all the loose grain remaining in the straw, and their construction allows the grain to fall through to the grain bottom.

The grain bottom, placed underneath the straw racks and the concaves, has a vibratory motion, and by this means the grain and the chaff are delivered to the inter-

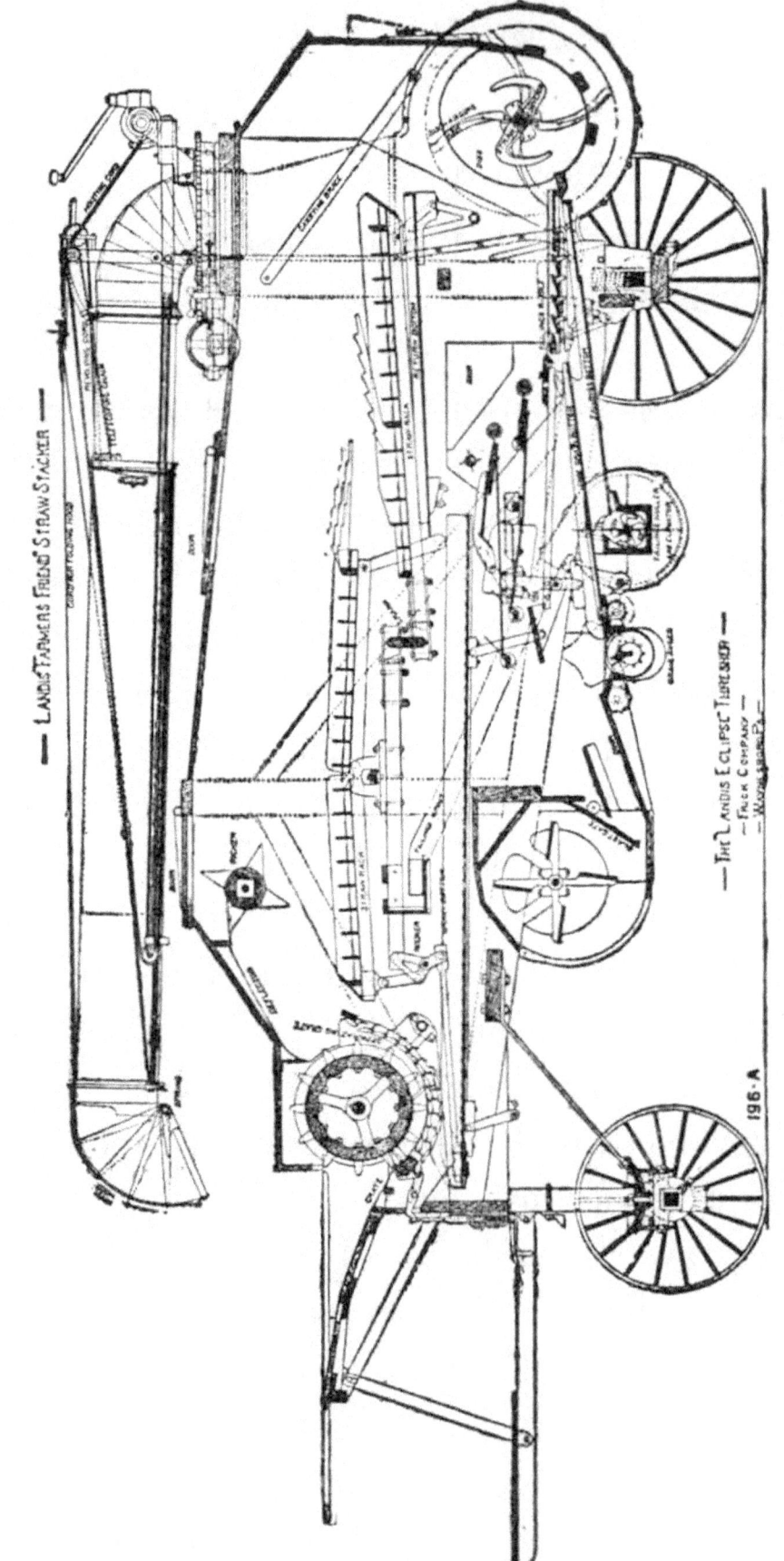

Fig. 66.—The "Landis Eclipse" Thresher.

mediate bottoms above the "shoegrain bottom," where the separation of the grain from the chaff takes place.

The chaff and tailings are, by the assistance of a blower or fan and by means of the "tailings riddle" and the "tailings bottom," carried to the "tailings huller" and elevator.

The grain falls through to the shoegrain bottom and from here to the grain auger.

The grain auger can deliver the grain to either side of the machine as wanted, or to other machines, such as a wagon loader, a weigher and bagger, etc.

The tailings are carried by the "tailings elevator" and a spout to a pan above the "grain bottom" or can be dumped under the machine, whence they can be removed.

The straw is taken care of by the stacker and deposited within twenty to thirty feet distance from the machine in a stack.

This gives you an idea of the general action of the thresher, and we will now describe in turn each individual part.

CYLINDER.

This important part has received much attention by the manufacturers. I give you below (see Fig. 67) a picture of the "Landis Eclipse Cylinder," and also (see Fig. 68) a picture of the Reeves' standard cylinder. As you can see, they differ somewhat in appearance. Both have teeth

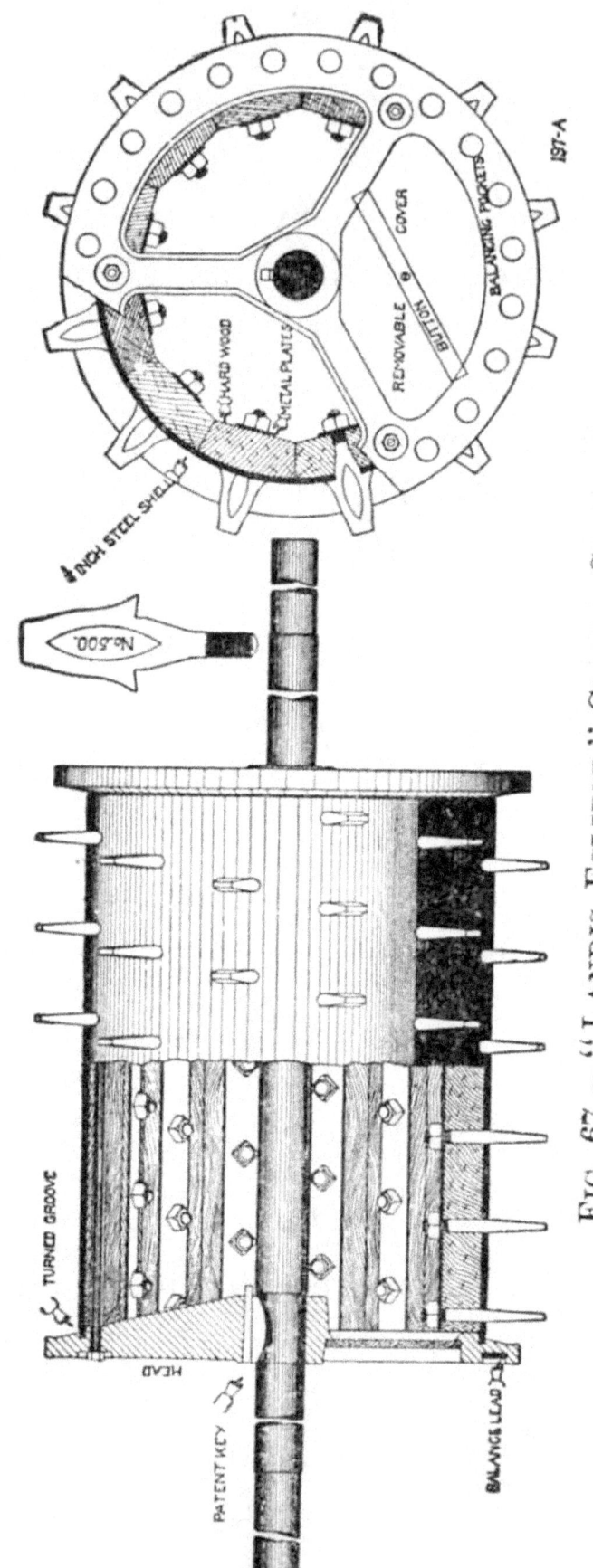

Fig. 67.—"Landis Eclipse" Closed Cylinder.

set in rows and both are built for high speed. The difference between them lies mostly in the fact that one is of a closed and the other of an open construction.

The closed cylinder has removable covers for the ends; this prevents the accumulation of dust and moisture inside, but does not hinder the adjustment or tightening of the nuts for securing the teeth. The claims for the open

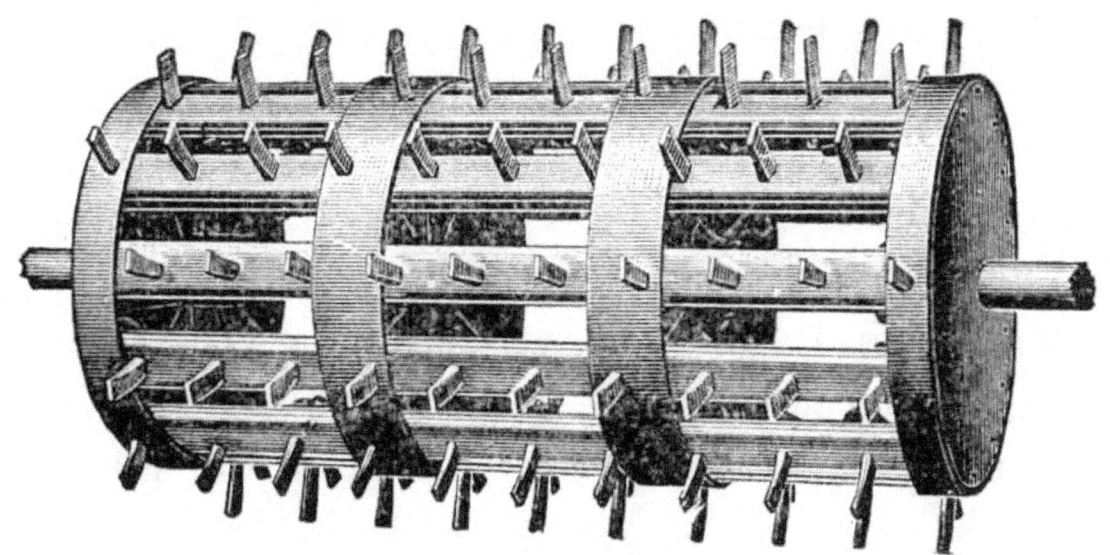

Fig. 68.—Reeves Standard Cylinder.

type are that the grain can fall right through as soon as it is loosened from the straw.

The shape of the teeth has also received much attention. If the front and the back of the teeth have the same curve, it follows that the cylinder can be turned end for end and that you will have practically twice the wear with only one cylinder setting.

As we pointed out before, the threshing is done by the cylinder teeth beating the straw while the concaves hold it in place. The adjustment of the position of the cylinder is therefore very important. It must run as near central as possible. If it runs too much towards one side, even if it does not strike the concave teeth it may still run close enough for cracking the grain. It is always advisable to allow a little end play, say about the thickness of a heavy wrapping paper on each end. This keeps the bearings in good order, makes the machine run easily, and saves power. Special care is taken to have the cylinder well balanced, as it runs at from 1100 to 1300 revolutions per minute. On account of this high speed, I want to caution you to be sure that the nuts, which hold the teeth, inside the cylinder are tight. With a new machine, it is advisable to try them every now and then, till you are sure that they are all tight. If a tooth gets bent or broken, stop down at once and replace it with a new one. If you do not have a spare one on hand, it is best to remove it and also to take out one on the opposite side, as this will keep the cylinder in balance. It is best to keep on hand a few extra teeth.

It must be remembered that in case your grain is tough or hard to thresh, you may have to increase the cylinder speed to 1200 or even 1300 revolutions per minute. This increase must, however, not be made by decreasing the diameter of your pulley on the cylinder shaft, because, as a rule, that is as small as can be allowed. Therefore speed up your engine until you get the correct number of revolutions you need for the cylinder.

I will say here that the larger the diameter of the pulley, the less is the slippage of your belt, and also that the faster your engine runs, the larger is the amount of power it develops. As it is also true that the faster the thresher runs the more power it requires, you can see the necessity of regulating the speed with the engine.

CONCAVES.

In the same cast-iron box on which the cylinder bearings are mounted are secured the concaves. These consist of plates, generally provided with two rows of teeth each.

The teeth are spaced so that, when properly adjusted, they come midway between the teeth of the cylinder. The concaves are removable and can be used in different places and in various numbers, depending on the kind of grain which is to be handled. They serve to hold the straw while the cylinder does the threshing. Different arrangements, both in regard to the number of teeth and position of them, must be used with different conditions of the straw. As a good rule you may put down in your memory that as few teeth as possible are to be used in your concaves. The less teeth you have, the less breaking or chopping of straw will occur and the less the power you will need, but do not forget that the first object is to get the grain out of the straw; therefore you must have enough teeth in the machine to thresh clean. Generally two rows will thresh oats; four rows wheat and barley; six rows flax and timothy.

The arrangement of the blanks and the rows of the teeth are of importance. When straw is dry and brittle, place a blank in front, as this will increase the "draw."

For "Turkey wheat" and "alfalfa" sometimes corrugated teeth are used. After a few experiments you will soon find out what is the best arrangement for each special case. Different arrangements for the adjustment of the concaves are used on different machines. A very good device consists of a series of levers regulated by a screw in the front of the machine; by this means the concaves can be moved nearer or farther away from the cylinder without changing their individual relations.

SEPARATING GRATE OR BEATER.

Immediately back of the concaves is placed the separating plate in the Eclipse machine, or the beater in the Reeves, the North West and others; the separating grate performing the same work as the beater. It consists of a grate with a curvature a little larger in diameter than the top of the teeth of the cylinder, and it extends a little

The Threshing Machine.

above the center of the same. This form allows the free discharge of the straw, but the grain and chaff pass through, the most of them separating from the straw before the straw leaves the concaves. The direction of the straw on leaving the grate is nearly vertical, and it is made to strike the deflector (a smooth steel plate) with the full velocity of the cylinder. Here most of the remaining grain is liberated and the straw is spread out over the full width of the separator in a thin sheet, allowing the grain to drop through to the grain bottom. At the end of the deflector sheet the straw picker (a revolving cylinder with triangular teeth) throws the straw down on the straw racks.

The other machines use, instead of the separating grate, a beater which revolves immediately back of the concaves. It consists generally of a cylinder with four or more blades of sheet-iron. The beater not only helps to separate the grain and the chaff from the straw, but it is also supposed to prevent the grain from flying around as well as to guide it to the grain bottom.

In combination with the beater is generally found a check board, which keeps the grain from being carried to the rear on top of the straw. In case the grain is damp or heavy there may be a tendency of the straw to stick to the cylinder; then the beater must be adjusted for more space and the check board must be raised so that the straw will pass freely.

STRAW RACKS.

The straw racks separate from the straw that grain and chaff which still remains after the previous operations have been finished. They allow the grain and the chaff to pass through, but not the straw. Owing to the oscillation of the racks, a strong up-and-down blast is produced between the openings. This motion and blast change the positions of the straw and the chaff, and the result is nearly a perfect separation.

The straw racks must have proper attention, because if they are out of order or not running properly, a great deal of grain is carried away with the straw. The less the straw is cut, the better the straw racks will work. It is therefore of importance not to use more teeth in the concaves than absolutely necessary for good threshing. All cranks and pitmans should have as little lost motion as possible and there should be no pounding. If the pitmans have worn short, new ones should be put in, and crank boxes should be moved forward by putting liners behind. The straw racks should be cleaned at least once a day, and all obstructions, such as roots, sticks, and stones must be removed, so that there is nothing to prevent the straw from going over and the grain and chaff from going through.

The Threshing Machine.

THE GRAIN CLEANER.

This is a new device found in the Eclipse, and I will quote from the manufacturer's description as follows:

"It is composed of three shelves (see Fig. 69); one below and in advance of the others. Under and along the front edge of the upper and middle shelves are supported, on guides, thin bars about four inches wide. These bars have right-angle notches along their front edge. At the front of each of these bars is a journaled roller, having right-angle corrugated grooves of the same depth and number as the notches in the bars. The rollers revolve in bearings which are pivotally supported upon a double rock-arm. This rock-arm has a quick vibrating motion which gives to the bearings and rollers a short, quick end-motion. The notched bars are also connected at one end to the roller bearings, and receive the same motion as the rollers. A convenient adjustment is provided on the outside of the machine for adjusting the notched bars to and from the rollers, for the purpose of increasing and diminishing the space between the notched bars and the rollers to suit the size of the grain or the amount of work to be done. This makes a cleaner that is quickly and conveniently changed while the machine is in operation, and does not require stopping of the machine to change from one kind of grain to another. The quick end-motion of the

258

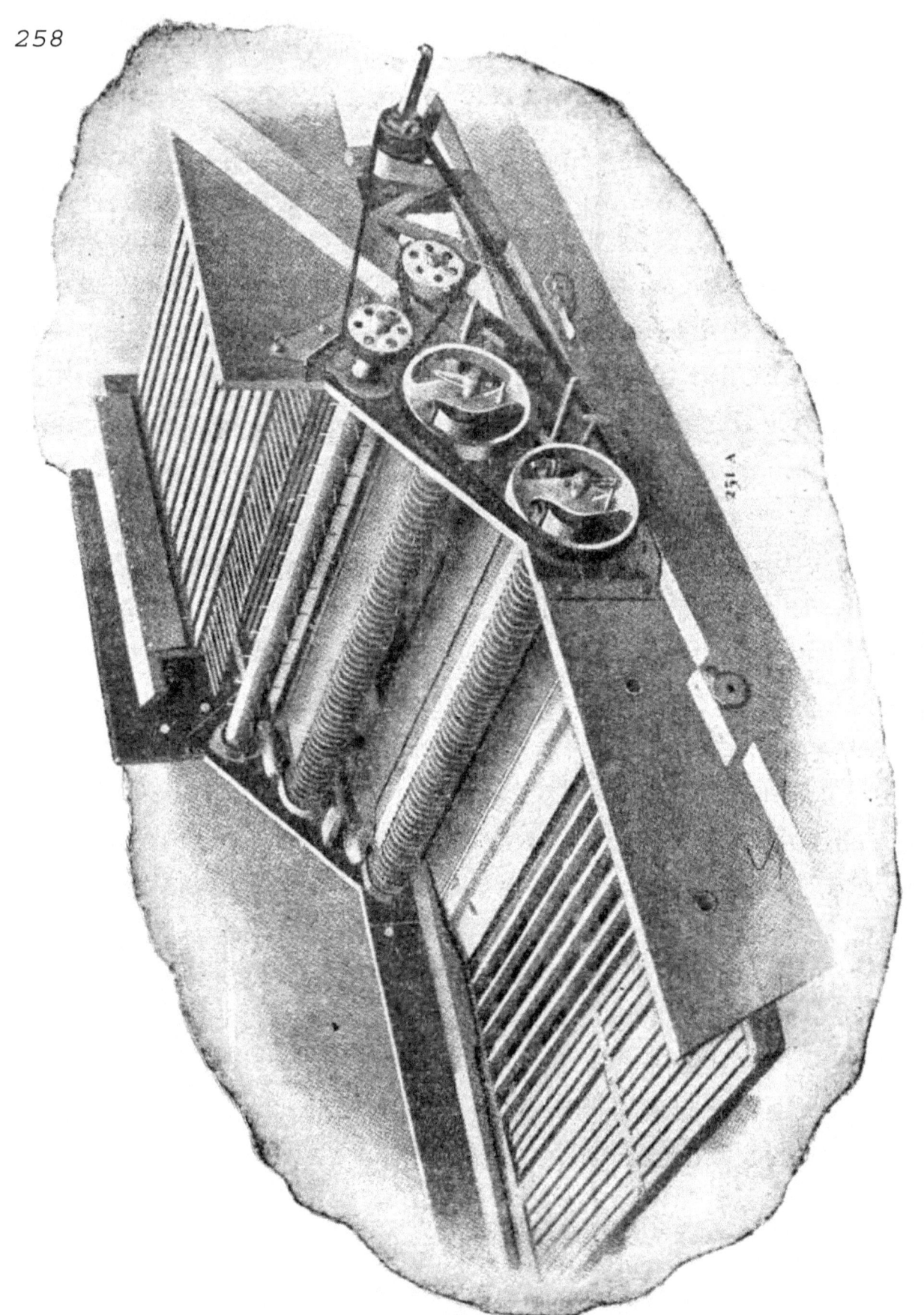

Fig. 69.

The Threshing Machine.

rollers and notched bars gives this cleaner great capacity with an opening of only sufficient size to let the grain pass through. This cleaner will separate all filth or foreign matter of less weight than the grain being cleaned, regardless of its size. To do perfect cleaning in timothy and flaxseed, a special attachment is required."

SIEVES AND SCREENS.

In most machines the cleaning is done by a number of sieves. The first of these, usually called the "conveyer sieve" or "grain bottom," is placed under the straw racks and conveys the grain to the final cleaning sieves, generally located in front of the fan. The proper selection of sieves, both in regard to mesh, number, and setting, is of great importance, and different sieves must be used not only for wheat, oats, barley, and others; but also for different kinds of the same grain. The setting of the sieves, as well as the rate of feeding, has much to do with the results. Close attention will soon show which is the proper sieve to use. Sometimes it may be necessary to use a screen in order to get rid of dust and small chaff. The screen is practically a sieve with very small mesh which does not allow the grain to pass through. Owing to the small size of the mesh, they very soon get clogged, and must then be taken out and cleaned. It is therefore not advisable to use them except when the grain is very dusty.

THE FAN.

The object of the fan is to supply a strong blast, directed against the current of the grain; it should blow away all chaff and dirt which has succeeded in passing by the cleaner or the sieves. For this reason, the strength of the blast must be carefully regulated, so that it will not affect the grain, but will remove all the lighter particles and the dirt. Every threshing machine has a regulator for this purpose, and the engineer must regulate the blast so that this result is obtained.

I will caution you here not to change the regulator too much each time you find that grain is blown over, as it is easy to pass by the point you want to find, that is, where the cleaning of the grain is accomplished without waste.

When the regulation is obtained by means of windboards, care must be taken that the board over the grain auger does not get bent, and it ought to be set so that the strongest part of the blast will strike the middle of the sieve. In very windy weather you must use a different adjustment than what you generally use or the result will not be the same.

Some machines use for this reason an equalizing blast-board which not only distributes the blast more evenly over or under the sieves as may be required, but also allows it to be concentrated in any direction. By adjust-

The Threshing Machine.

ing this board carefully, the effect of the wind can be neutralized.

THE GRAIN AUGER.

After the grain has reached the "shoegrain bottom" it drops into the grain auger, which conveys it to either side of the machine, or to such additional machinery as may be wanted. It consists simply of a long pitch screw revolving in a box. By reversing the belt or placing the drive chain on top or bottom on the sprocket wheel on the shaft, the grain auger will deliver the grain on the opposite side of the machine.

THE TAILINGS AUGER.

Underneath the grain bottom is the tailings bottom. This delivers the chaff and tailings to the tailings auger and huller. A centrifugal elevator delivers the hulled tailings on the first grain bottom. A door in the bottom of the huller allows the tailings to be delivered on the ground. This is a very good way to dispose of the tailings, and does not allow bolts or nuts or small tools to get into the cylinder. The tailings are usually carried back to the cylinder by an elevator driven by a chain. This chain must not be too tight or it will take too much power, but must be tight enough not to ride over the sprocket wheel.

If the chain does run off, the best way to replace it is to tie a weight to a rope and drop it down through lower part of the elevator to a man underneath the machine. Tie the end of the chain to the end of the rope and let the man on top pull it up. The chain is fed in underneath by the man, who must see that it has no kinks and that it is fed in straight. When the chain has reached the man on top of the machine, the weight is dropped down again in the upper portion of the elevator, and the man under the machine must pull down on the rope, at the same time feeding in on the chain until he has both ends. After hooking the chain together, put it on the lower sprocket first, and then on the upper, and take up the slack with the adjusting set screws on top. Turn the chain around once or twice to see if it is straight and if the adjustment is right.

Tailings should be small and contain very little grain. If too much tailings are returned it may choke the cylinder, and besides there is always danger that a great deal of the grain so returned may get cracked in the cylinder. It is best to keep the returned tailings as low as possible.

THE STACKER.

After the straw leaves the straw-racks it is taken care of by the stacker. The ordinary stacker is simply an endless chain of slats which carries the straw up an incline and drops it on top of the stack. Lately a new apparatus for taking care of the straw is coming into use under the name of the "wind stacker." This consists of a fan furnishing a strong current of air which carries the straw through a tube and places it on top of the stack. Various kinds of wind stackers are used. The one shown in our figure 66, the Landis "Farmer's Friend," consists of a straw-drum into which the straw and the chaff are delivered from the straw racks and tailings riddle. In this drum, on the same shaft as the fan, revolve a number of curved fingers, which take the straw when it is falling from the straw racks and give it a revolving motion towards the inlet of the stacker fan.

The fan and the straw are now revolving in the same direction and with nearly the same speed. The result is that the straw passes through the fan without breaking. The other side of the fan being closed, the air, having passed through the fan, cannot return, and must therefore continue out through the delivery pipe, carrying the straw with it. In order to allow a compact bundle of straw to pass by without stopping the fan, the fan shaft has a relatively large end-play which enables the fan to move sideways so that the bundle can pass and to return to its proper place as soon as it has passed through. The straw pipe can be telescoped and is oscillated automatically at any part of the circle. The automatic device can conveniently and quickly be disconnected and the stacker directed by hand to any position wanted, and can then again be connected to the automatic, which will then give to the pipe the same motion as before. The straw pipe stops automatically a few seconds before reversing at each end of the stroke. This keeps the ends of the stack and the middle at about the same height. At the end of the delivery pipe is a deflecting nozzle which further directs the straw in any desired direction relatively to the delivery pipe. As the telescoping pipe can also be made to revolve round itself, the straw can be thrown at any angle of a circle. The device can therefore be used for packing straw into any corner of a barn or loft as well as in making any shape of a stack.

Most of the wind stackers are similar to the one described, but each manufacturer's design is somewhat different in detail. The "Fosston" wind stacker makes use of a vertical fan and does not carry the straw through the fan, which is somewhat of an advantage.

SELF-FEEDER AND BAND-CUTTER.

This is a machine designed not only to carry the straw bundles up and deliver them to the cylinder, but also to cut the band of the bundles, and to distribute the straw under the cylinder after the bands have been cut.

The straw carrier is arranged in the same manner as

The Threshing Machine.

the straw carrier in the stackers and delivers the straw bundles under the band cutters. It also holds them in position while their bands are being cut. The cutter consists of a number of cutting arms, each one of which holds three or more sickle-edged cutting sections. The cutting arms are adjustable to the size of the bundles. Back of them, attached to the wooden bars, are the lifting blades, which are also sickle-edged. Any band which should have escaped the first set of knives will be cut by the second. The knives lift also the butt end of the straw bundles, while the heads of the same are depressed to meet the cylinder teeth by pronged forks placed at the extreme end of the wooden bars. A separate governor regulates these devices and prevents uneven distribution of the straw fed to the cylinder, as well as all crowding. Probably the "Parsons" machine is the one most used. It has no complicated parts and is very sensitive. The feeder is driven from the cylinder pulley on the thresher by a wide belt running over the pulley on its crank shank. Generally a small belt tightner is provided on the other end of the crank shaft. A pair of bevel gears run on inclined shafts from which the governor receives its motion by means of another pair of bevel gears. The governor is of the regular centrifugal type and regulates by engaging or disengaging the feed shaft.

The carrier is also driven from the inclined shaft by

means of a friction wheel and a disc. By moving the friction wheel out or in on the inclined shaft any desired speed can be given to the carrier. A crank is supplied for this purpose, and the speed can therefore be changed at any time without stopping the machine.

Attachments.—In combination with the thresher, various attachments are used. Generally a weigher, a bagger, and a loader constitute the rig. To these are sometimes added a "Nesmith" or a "Miller" grain-register, a dust collector, or a pneumatic grain elevator.

PART FOURTEENTH.

HOW TO RUN A THRESHING RIG.

Before you uncouple your traction engine go over the threshing place carefully; make note of the slope of the ground and the direction of the wind. The best position you can secure relatively to the wind is when the straw on the road to the stack moves a little to one side but in the same general direction as the wind.

See that you have room for your stack, that loaded and unloaded wagons can pass, and that nothing can interfere with your belts. Bearing these points in mind, select the position that will be most convenient and haul your thresher in place. If the ground is not level, dig holes for the wheels which are too high and block the rear wheels well. It is best to carry a level with you, as this saves time in the end. Be careful that the machine is level crosswise. It is not so necessary that it should be level lengthwise. In fact, it is of some advantage to have the cylinder end four to six inches higher the cleaner end;

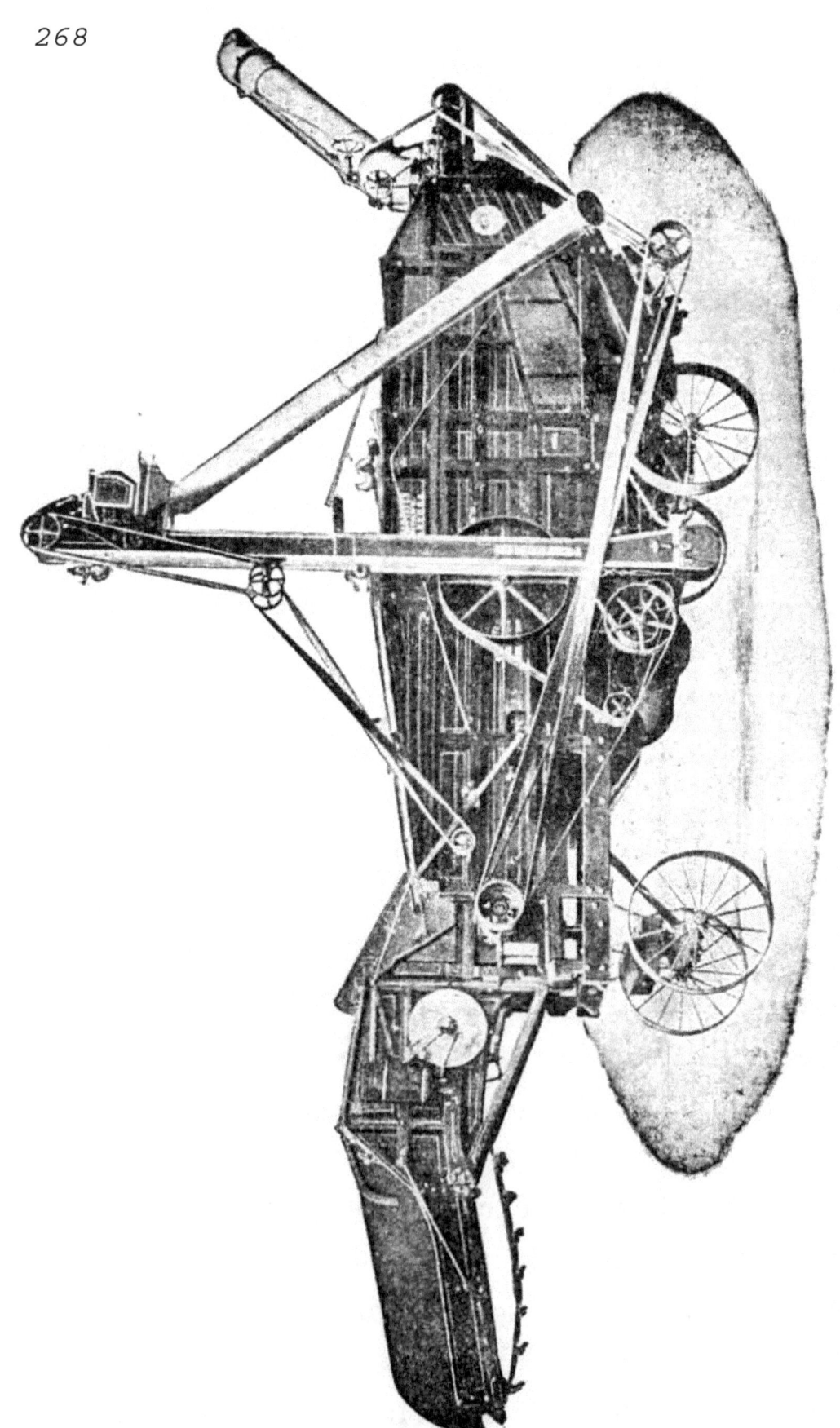

Fig. 70.—Buffalo Pitts Niagara Second Thresher.—Fitted with Buffalo Pitts Wind Stacker, Parsons Feeder, and Perfection Weigher and Wagon Loader (Dakota Style).

but remember it must not lean the other way. See that the blocks of the right hind wheel are tight, so that the pulling of the belt will not disturb the setting. If there are jack-screws above the front and rear axles, to make the four corners solid adjust them and screw the nuts down tight. Now you can uncouple and pull your traction engine to such a position that the driving-wheel and the pulley on the thresher are in line. Take your time and do this right. It is better to spend a little more time now than to have your belt run off when you are running with a heavy load on the machine. Most of the pulleys are crowned, that is, they are a little larger in diameter in the center than at the sides, and the tendency of the belt is to run in the center; but if the machines are not in line and the shafts not horizontal, the belt will nevertheless run off. Before you put the belt on, go over the machine carefully and clean all the bearings and oil holes well. Take off the belt-tightner pulley and clean the oil chambers and the spindle, oil them and put them back. If the machine is new, more care must be used when doing this, as paint may have got into the bearings and oil holes, in which case it is best to remove the shaft and carefully scrape the paint off. Wipe out the bearing and the oil hole with clean waste, oil them well, and replace the shaft. After you are sure they are all cleaned, put a few drops of oil in each oil hole. Use only No. 1

machine oil with good body. If the machine uses grease, see that it is good. Don't use axle grease, as it very often contains resin, which will deposit on the bearings and cause them to heat.

Now put on the main belt, and be sure that the machine runs the right way. You can easily see if the cylinder will pull the straw in or not. If it should run the wrong way, cross your belt. I will say here that it is not good practice to run with a crossed belt, as it takes more power. Sometimes, if the belt is short, and the difference in diameter between the pulleys is great, a crossed belt will give more power, but it is better not to run it crossed except when it is necessary.

Another thing I want you to bear in mind is, that you must always run your leather belts, if they are single, with the hair side on the pulley. They run better and wear better. The reason is that the flesh side is more flexible than the hair side, and therefore does not crack so easily. When a belt travels over a pulley, the outside is longer than the inside, and the difference is made up by stretching, which the flesh side is better able to stand than the hair side. If you use double belts, it does not matter which side you put against the pulley, as both sides are alike.

As we are talking of belts, I will also caution you not to use more tension than is needed to prevent slipping, as this will not only hurt your belts, but also the bearings. You will also find that if you remove your belts as you stop down at night, or when your threshing is done, they will last longer. Rubber belts must run with the seam on the outside.

After the belt is on the cylinder pulley, see that the oil cups are feeding or that the grease cups are full and screwed down and that no other belt is in place. Turn the cylinder over by hand a few turns to find out if it is properly adjusted and that everything is in place, and see that no tools or bolts have been left about where they can get into the machine. Now start up and let the cylinder run. Keep an eye on the bearings to see if they heat. Go over the next shaft the same way, and when this is ready to start, stop down and put your belt on. Turn it over by hand first to see that it does not bind or that the bearings are loose. Try all your bolts and nuts, and if they are loose, tighten them before you put on the belt. After you have put the belt in place, start up again and get the next shaft ready, and so on until all your belts are on and running. As you are putting each belt in its place, note carefully if it has to run straight or crossed, or, in other words, see that the shaft it is to drive runs in the direction which it should in order to make the thresher do its work. When you first start up, is the time to give your machine all the oil it needs, and as a rule it is well to give each bearing an additional drop or two of oil after the machine is running, as the oil then will distribute itself better. After you have started to thresh, it is well to oil freely for the first two days, to be sure all dust, dirt, and grit have disappeared, after which time less oil is needed. Before you put belts on a pulley be sure that the key or the set screws are not only in their places, but that they are tight. If the machine has self-lining bearings, see that they are not bound in their seats but can move and are free enough to adjust themselves to the shaft. After attending to all this, it is best to let your machine run for a little while empty. This gives you an opportunity to put your speed counter on the cylinder shaft and find out how fast it is running. If you find that it makes less than 1100 revolutions per minute, increase the speed of your engine until the cylinder reaches this speed. If you do not have a speed counter, take the speed at the engine. To do this, place your left hand on the projecting end of the crank-shaft, and holding a watch —which has a second hand—in your right hand, you can count the number of revolutions the engine makes in one minute. If the diameter of the driving wheel on the engine is 36 inches, and the diameter of the pulley on the cylinder shaft is 9 inches, you know that when the large pulley makes one revolution, the small one makes four. If you have only an eight-inch pulley on the cylinder shaft, the small one will make $4\frac{3}{8}$ revolutions for each revolution of the large one. If you counted 250 revolutions per minute as the speed of the engine, the speed of the cylinder would be in the last case $250 \times 4\frac{3}{8}$, or 1125 revolutions per minute.

Now examine your straw and make up your mind how many concaves you will need, as well as make a first setting of your cleaner. Be sure that you have enough space between the notched edges of the shelves and the grooved rollers to allow the grain to pass through. Remember, also, that it is impossible to get a good separation without clean threshing, and for this reason it is important to pay careful attention to your cylinder and your concaves. Also take a last look at the straw racks, so that you are sure that there are no obstructions left on them. Now send enough straw through the machine so that you can judge how it will work. As soon as the grain has passed from the spout, examine it carefully, and if you find that it contains small and light dirt, it is clear that you have not given it enough blast, and therefore you must move the short lever on the blast gate down a notch. Now examine again, and if still not clean, move your lever down another notch, and so on until you find the right place. If you should find large as well as small dirt in the grain, then the spaces between the rollers and the shelves in the cleaner are too large, and must be adjusted. If the grain is clean, then examine the tailings and see how much grain appears here. If you find a great deal of grain, it indicates that the spaces between rollers and shelves are too small for the amount going through the machine. The remedy is to make them a little larger, but it is possible you may have too much blast. You will soon be able to decide which is the cause. After the machine has been adjusted so that everything seems all right, it is best to examine the straw as delivered from the stacker. If you should still find grain in the straw, it shows that you are not threshing clean. The trouble is then in the adjustment of the concaves, or that you have not enough of them in. With the Landis Eclipse it is generally enough to use only one concave in front and adjusted close to the cylinder; but if you find that you are not threshing clean, place one solid bank in front and a concave in second place, or you may use two concaves, one in first place and another in second place. If the grain is tough, or what is known as "headed grain," use both concaves with a solid ribbed concave between them. If you find the "headed grain" does not thresh clean with this arrangement, it is best to increase the speed of your cylinder. From 1200 to 1300 revolutions a minute may have to be used.

After these adjustments are all made, increase the feeding of the straw so that the amount is about normal and take the speed of the cylinder shaft again. If you find that it is making a less number of revolutions than before, increase the speed of the engine till again you have the right number. Examine your grain and tailings again to see if the increase of speed has changed the results. If they have, you must keep on adjusting until you get the result you want. If your machine has a self-feeder attached, now is the time to get it started. It is not advisable to let this part run when you are adjusting the rest of the machine, or before the thresher is running at regular speed. A few sheaves laid on it will prevent it from moving. The best way, however, is to keep the belt off the pulley. It is easy to put it on when you need it. Before putting the belt on, see that the knife arms are adjusted to the size of the sheaves and the condition of the bands. It is taken for granted that you have taken the same care of this part of the machine as of the others in regard to bearings adjustment, cleaning, and oiling. The bearings and cranks are to be oiled regularly; but after the break-band has been run in and is smooth no more oil is needed. If you are using an attached stacker, it is well to remember that you must not use the automatic moving apparatus and the hand device at the same time. If you want to operate it by hand, it is best to take the belt off the automatic device. The speed of the stacker should be adjusted to the work it has to do by changing the pulleys. The faster it runs, the more power is required, therefore, keep your speed as low as possible. If less speed is required put on a larger pulley.

If you are using a wind stacker see that your fan is running all right and that it has enough end-play. In case of the Landis Farmer's Friend, this must not be less than $\frac{5}{8}$ of an inch, and can be one inch. The hub of the straw propeller must be 8 inches from the hub of the fan. This stacker has the automatic reversing gear located on top of the thresher convenient for the operator. It can be connected and disconnected very quickly and the straw pipe can be stopped in any position desired by simply pushing the small lever to the center position and dropping the pawl into the notch on the cover. This operation disconnects the automatic gear. If it is desired to run the pipe by hand, it is best to take off the belt on the automatic device and loosen the locknut pin on the pinion.

Before you move the machine, see that the pipe is put in its bracket and well fastened. In case of fire it may become necessary to move your machine quickly, and under those conditions remove the blocks under the wheels, put a man at the tongue for steering, throw in your friction clutch on the traction engine, and back out slowly, pulling the thresher out by means of the main belt. A few men at the wheels will be of great help for starting and may save time. If hard to get started, let the man on the tongue steer the front wheels clear around. This saves often a lot of time.

There is no need of telling you that you must see that boxes and shafts are properly oiled, just as on all other machinery.

In most threshers the boxes are babbitted, and if they should get hot are liable to melt out. In such a case the only remedy is to rebabbit. In order to do this, you need preferably a plumber's furnace, a good-sized ladle, and some good babbit; as some boxes take three to four pounds of babbit, the ladle ought to hold at least five to six pounds of melted metal. The box to be babbitted must have all the old babbit removed and the surface must be cleaned carefully with waste to remove all dirt and as much grease as possible. Having done this, it is best to wash it with gasoline to remove such grease or oil as could not be reached by the waste. It is important to get all the oil or grease removed from the surface you want to babbit, as it not only prevents the babbit from sticking to the box, but also forms gases producing blowholes in the babbit, which may compel you to do the work over again. When you are sure that the box is cleaned, you must clean the shaft and wrap around it a sheet of good paper. A little glue on the edges, which ought to overlap each other, will hold it in place. See that you put the paper on smooth and even and that it is a little longer than the box, so that it protrudes over the ends. If the box you are rebabbitting is solid, place it over the paper on the shaft and block it up, so that the space all around between the box and the shaft is the same and equal on both ends, or, in other words, see that the shaft is central in the box. Now secure some good stiff clay or putty and close up the ends with it, leaving a hole on the top at each end for the air to escape. Make a funnel around the oil hole on top to pour the hot metal in. If there is more than one hole in the top of the casting, put a wood plug through the oil hole to the shaft and make funnels for pouring over the other or others, as in that case you will not have to drill out the oil hole after the box is babbitted. Now see that your babbit is hot enough. If a stick of white pine is charred when put into the ladle the temperature is right. You can now pour your metal. Keep it running in an even stream until the metal flows out of the top of the air holes on the ends. When you have begun pouring, don't stop if some runs over; don't let that trouble you—keep on pouring till the box is full. After the babbit is cooled off, remove the box from the shaft, clean off the clay or putty, remove the wood plug in the oil hole and trim off the edges. Cut a couple of slanting grooves in the top of the box with a half-round cold chisel, starting at the oil hole and ending near the ends and close to the bottom.

How to Run a Threshing Rig.

If the box you had to rebabbit had been of the "split" kind you would have handled it in the same way, except that you would have had to get a couple of pieces of cardboard or sheet-iron and place them between the top and the bottom casting. Make them wide enough, so that they rest tight against the shaft, and cut holes for the bolts, which hold the two halves of the box together, to go through.

Put enough of those cardboard liners between the two halves so that you can take up on the box when it wears down. To allow the babbit to run through from the top to the bottom of the box, cut notches in the edges of the cardboard liners resting against the shaft. Make the notches about $\frac{1}{4}$ to $\frac{3}{8}$ of an inch deep and about one to one and one-half inches apart. When the babbit is cold, loosen up on the bolts a little and drive a sharp cold chisel in between the two halves of box to break them apart.

After you have separated the top and the bottom, take a rough file and smooth down the edges, and with a sharp chisel cut off the pouring gates and also cut in your oil grooves. All this being done, take off the paper from the shaft. If the shaft should be rough, smooth it down with emery cloth or paper and then put a little oil over the surface. You can now replace the bearing on the machine.

As we have seen, a threshing machine makes use of a

great number of belts, and its running well depends to a great extent on how you handle and take care of them. We have pointed out several times how they ought to be used. It is to be remembered that they should not be allowed to get wet, and if they get dry and hard they must be softened by rubbing into them a little "neat's foot" oil. A soft belt transmits more power than a hard one and does not wear nearly as much.

The amount of power transmitted through a single belt can be roughly estimated by the following rule, which it is easy to remember: one inch belt running with a velocity of 800 feet per minute will transmit about one horsepower. In order to apply this rule, you will have to know the diameter of the pulley over which the belt runs, and the number of revolutions it makes per minute. If the diameter of the pulley is given in inches, you must divide by 12 to reduce it to feet, and then multiply by 3.1 in order to get the circumference of the pulley in feet. Multiply this sum by the number of revolutions of the pulley per minute, and divide by 800. The result is the number of horsepower each inch of belt can safely transmit. By multiplying this by the width of the belt, the total amount of power you can transmit through this belt is ascertained. For example, if your pulley is 8 inches in diameter and makes 1200 revolutions per minute, how many horsepower can an 8-inch wide single belt transmit? The calculation is made as follows: diameter of pulley is 8 inches, which, divided by 12 inches, gives $\frac{2}{3}$ of a foot. This multiplied by 3.1 to find the circumference gives 2.05 feet. As the pulley makes 1200 revolutions per minute, we must multiply 2.05 by 1200, which gives us 2460 feet per minute. Dividing this sum by 800, we find 3.75 horsepower per inch as answer. Now multiplying 3.75 by 8, the width of the belt, will give us 30 as answer, or we can say that we can safely transmit 30 horsepower under those conditions.

In order to get the best results out of the belt with the least strain on the bearings, it is advisable wherever possible to arrange matters so that the slack side of the belt is on top and the pulling side underneath, as in this condition the slippage is materially decreased. The belts ought to be examined every day to see that they are in good order. If any lacing shows signs of coming loose, relace as soon as possible.

Lacing should be done so that the joint is as nearly like the rest of the belt as possible. The lacing should always run in the same direction as the belt on the pulling side and be crossed on the outside. When you want to lace a belt, put a square against the side and cut it off. Then mark your holes evenly about an inch apart; which is a good practice, and see that you get a hole from $\frac{1}{4}$ to $\frac{3}{8}$ of an inch from each edge. After the holes are marked, use a good sharp belt punch for cutting them. If you are using a hand-punch and a hammer, rest your belt on the end grain of a block of wood and cut through with one blow if possible. If the belts run over very small pulleys, some engineers use a hinged lacing, that is, passing the lacing from one hole around the end of the belt and down through the next hole on the other side of opposite end of the belt. This way of lacing is not so good as the standard one, and is only to be used in special cases. When the lacing is finished, fasten the ends by making a narrow slit with a knife on the place where one hole will come when you have to cut your belt next time, and force your lace through this slit. After the lace is through, cut it half-way across close to the surface of the belt and cut it off about $\frac{1}{2}$ inch from the belt. By twisting the half-inch long piece around so that it is at right angles with the rest of the lace, it will not slip through. If you are using cotton belting, make the holes with an awl instead of with the punch, as it does not hurt the fibres so much.

Next in importance to the belts is the cylinder. As we have seen, it makes about 1100 revolutions per minute. If we suppose that the diameter of the cylinder is 30 inches, the circumference is about 30×3.1 inches, or 93 inches. If we reduce this to feet by dividing by 12 inches, we find it equal to 7 feet 9 inches. As the cylinder is supposed to make 1100 revolutions per minute, the surface of the cylinder travels $7\frac{3}{4} \times 1100$ feet, or 8,525 feet per minute. If you now divide this sum by 5,280, which is the number of feet in one English mile, and multiply by 60 (minutes per hour), you find the surface of your cylinder travels at the rate of 96.86 miles per hour, or considerably faster than an express train. I wanted to call attention to this fact in order to impress on your mind how important it is to have your cylinder well balanced if you expect it to run well. It also explains why I advised you to replace any tooth which had been bent or had to be removed from the cylinder for other causes. If you should disregard this rule, you would soon find that your machine would vibrate very much and try to move all over the place. This would be due to the fact that the cylinder would be out of balance. If in this condition the run is kept up very long, you would soon shake all the bolts in your machine loose, wear out your bearings, and perhaps bring the shafts out of line—a very serious trouble. Even with all the teeth in place the cylinder may become unbalanced. If there is any doubt in your mind in regard to the balance of the cylinder, it is best to test it at your earliest opportunity. The best way to accomplish this is to remove the cylinder from the thresher and to place it between two or more heavy wood blocks with a smooth surface, so that it rests only on the journals. Put a spirit level on the cylinder shaft and drive wedges under the blocks till it is level. See that the wood blocks are also level, so that the cylinder has no tendency to roll one way or the other. Secure two straight pieces of steel, square or round, and place them under the journals and on top of the wood blocks. Measure the distance between their ends, and if it is not the same make it so. You are now ready for your test. Start the cylinder rolling a little, and when it comes to rest, mark with chalk the highest point on the cylinder. Do this two or three times and make the same kind of marks. If the mark comes nearly at the same spot, you can conclude that your cylinder is out of balance and that the spot marked is the light side. In case the cylinder is of the enclosed type you will find that it has small pockets in each end. The object of them is to provide a place where the weights needed for balancing can be placed. After you have marked the light side of the cylinder place additional weights in the side pockets in line with this mark and try the cylinder again for balance. If it comes to rest on the same spot, it proves that you have not put on enough weight, and if it stops opposite the old mark, you have put on too much. After a few trials you will find a weight which will allow the cylinder to come to rest at any point, and that is the result you want to reach, as that is the proof that the cylinder is balanced.

If the cylinder is of the open type no pockets are provided and you will have to balance it by driving small wedges under the centre band at the mark. This operation must be continued as long as the cylinder comes to rest at the same place after each trial.

It is hardly necessary to point out that it is best to put your thresher in first-class order after the season is over, and before you put it away for the winter. Every part of the machine ought to be cleaned, and when needed repaired; boxes ought to be looked over and rebabbitted if worn, all nuts and bolts retightened and replaced if lost, all teeth in cylinder or concaves which are bent or worn out replaced with new ones, and eccentrics and pitmans gone over and repaired. Every two or three years it will need a new coat of paint. When it is in good order, cover it with canvas and store it in a dry place. If treated in this manner the thresher will last for a great many years.

THE FEEDER.

If you are not an expert feeder, you will have to start slowly and pay attention to the machine. To become a good feeder takes long practice. A good feeder keeps the straw carrier evenly covered with straw, and keeps also an eye on all the rest of the machine, so that he knows that the machine is doing and can quickly correct anything which goes wrong. The stacker and the tailings are sure indicators of how the machine works. The cylinder ought to be kept full all over its surface and the bundles of straw should be tipped well up against the cylinder. A bundle is easy to spread and one on each side of the machine can be handled. Flat bundles should be fed on edge.

It is advisable for the feeder to examine the forks used by the pitcher in order to see that they are tight on their handles, as a loose fork in a thresher is a very serious thing. Never let any one be used which is not absolutely tight on the handle.

Weight per Bushel of Grain.

WEIGHT PER BUSHEL OF GRAIN.

The following table gives the number of pounds per bushel required by law or custom in the sale of grain in the several States:

	Barley.	Beans.	Buckwheat.	Clover.	Flax.	Oats.	Rye.	Shelled Corn.	Timothy.	Wheat.
Arkansas	48	60	52	60	..	..	56	56	45	60
California	50	..	40	..	..	32	54	52	..	60
Connecticut	..	..	45	..	..	32	56	56	..	56
District of Columbia	47	62	48	60	..	32	56	56	45	60
Georgia	40	..	..	60	..	35	56	56	45	60
Illinois	48	60	52	60	56	32	56	56	..	60
Indiana	48	60	50	60	..	32	56	56	45	60
Iowa	48	60	52	60	56	32	56	56	45	60
Kansas	50	60	50	..	..	32	56	56	45	60
Kentucky	48	60	52	60	56	32	56	56	45	60
Louisiana	32	..	..	..	..	32	..	56	..	60
Maine	48	64	48	..	..	30	..	56	..	60
Maryland	48	64	48	..	..	32	56	56	45	60
Massachusetts	48	48	..	..	..	32	56	56	..	60
Michigan	48	..	48	60	56	32	56	56	45	60
Minnesota	48	60	42	60	..	32	56	56	..	60
Missouri	48	60	52	60	56	32	56	56	45	60
Nebraska	48	60	52	60	..	34	56	56	45	60
New York	48	62	48	60	..	32	56	58	44	60
New Jersey	48	..	50	64	..	30	56	56	..	60
New Hampshire	..	60	..	..	..	30	56	56	..	60
North Carolina	48	..	50	64	..	30	56	54	..	60
North Dakota	48	..	42	60	56	32	56	56	..	60
Ohio	48	60	50	60	..	32	50	56	45	60
Oklahoma	48	..	42	60	56	32	56	56	..	60
Oregon	46	..	42	60	..	36	56	56	..	60
Pennsylvania	47	..	48	62	..	30	56	56	..	60
South Dakota	48	..	52	60	56	32	56	56	..	60
South Carolina	48	60	56	60	..	33	56	56	..	60
Vermont	48	64	48	..	60	32	56	56	42	60
Virginia	48	60	48	64	..	32	56	56	45	60
West Virginia	48	60	52	60	..	32	56	56	45	60
Wisconsin	48	..	48	60	..	32	56	56	..	60

288

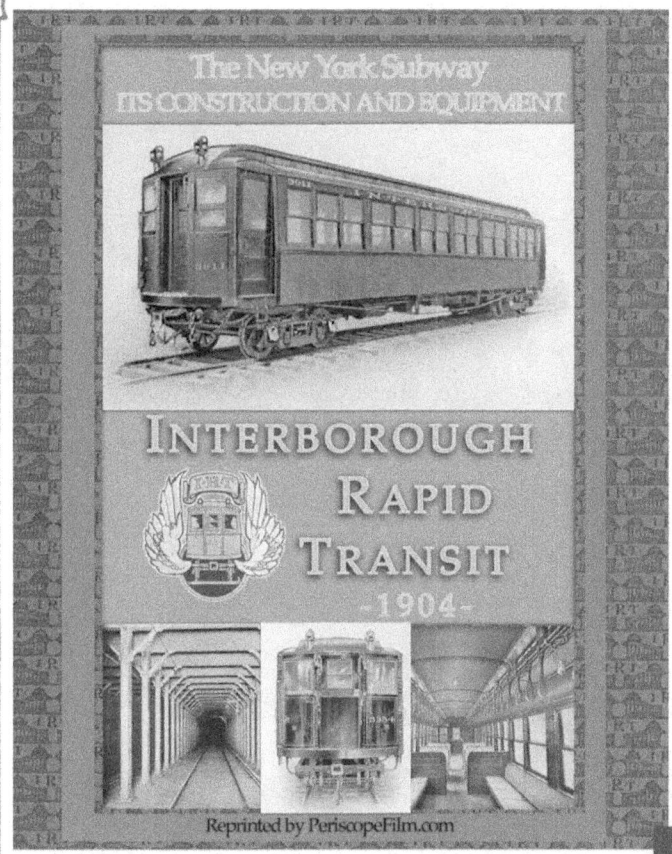

On October 27, 1904, the Interborough Rapid Transit Company opened the first subway in New York City. Running between City Hall and 145th Street at Broadway, the line was greeted with enthusiasm and, in some circles, trepidation. Created under the supervision of Chief Engineer S.L.F. Deyo, the arrival of the IRT foreshadowed the end of the "elevated" transit era on the island of Manhattan. The subway proved such a success that the IRT Co. soon achieved a monopoly on New York public transit. In 1940 the IRT and its rival the BMT were taken over by the City of New York. Today, the IRT subway lines still exist, primarily in Manhattan where they are operated as the "A Division" of the subway. Reprinted here is a special book created by the IRT, recounting the design and construction of the fledgling subway system. Originally created in 1904, it presents the IRT story with a flourish, and with numerous fascinating illustrations and rare photographs.

Originally written in the late 1900's and then periodically revised, A History of the Baldwin Locomotive Works chronicles the origins and growth of one of America's greatest industrial-era corporations. Founded in the early 1830's by Philadelphia jeweler Matthais Baldwin, the company built a huge number of steam locomotives before ceasing production in 1949. These included the 4-4-0 American type, 2-8-2 Mikado and 2-8-0 Consolidation. Hit hard by the loss of the steam engine market, Baldwin soldiered on for a brief while, producing electric and diesel engines. General Electric's dominance of the market proved too much, and Baldwin finally closed its doors in 1956. By that time over 70,500 Baldwin locomotives had been produced. This high quality reprint of the official company history dates from 1920. The book has been slightly reformatted, but care has been taken to preserve the integrity of the text.

NOW AVAILABLE AT
WWW.PERISCOPEFILM.COM

©2008-2011 Periscope Film LLC
All rights Reserved
ISBN #978-1-935700-56-2

www.PeriscopeFilm.com

www.ingramcontent.com/pod-product-compliance
Lightning Source LLC
Chambersburg PA
CBHW082110230426
43671CB00015B/2663